THE MYSTERY OF MATTER

The Mystery of Matter

Jennifer Trusted

First published in Great Britain 1999 by
MACMILLAN PRESS LTD
Houndmills, Basingstoke, Hampshire RG21 6XS and London
Companies and representatives throughout the world

A catalogue record for this book is available from the British Library.

ISBN 0–333–76002–6

 First published in the United States of America 1999 by
ST. MARTIN'S PRESS, INC.,
Scholarly and Reference Division,
175 Fifth Avenue, New York, N.Y. 10010

ISBN 0–312–22145–2

Library of Congress Cataloging-in-Publication Data
Trusted, Jennifer.
The mystery of matter / Jennifer Trusted.
p.   cm.
Includes bibliographical references and index.
ISBN 0–312–22145–2 (cloth)
1. Matter.   2. Substance (Philosophy)   3. Metaphysics.   I. Title.
QC171.2.T78   1999
117—dc21   98–54238
                                                                    CIP

This book is printed on paper suitable for recycling and made from fully managed and sustained forest sources.

10   9   8   7   6   5   4   3   2   1
08   07   06   05   04   03   02   01   00   99

Printed and bound in Great Britain by
Antony Rowe Ltd, Chippenham, Wiltshire

Contents

Acknowledgements

My grateful thanks for very helpful advice from Professor D.J. O'Connor and from Professor Richard Seaford in regard to classical Greek writers and to Dr Trevor Preist who has been good enough to read through my account of developments in twentieth-century physics. I am also indebted to material written by Russell Stannard and Noel Coley in the Open University course A382, Modern Physics and Problems of Knowledge. In addition I should like to thank Mr Martin Davies for kindly reading through the manuscript and for invaluable suggestions as to explication and style.

Preface
Why is Matter a Mystery?

The word matter is, in philosophy, the name of a problem.[1]

What can be more self-evident and certain than the existence of matter? There may be doubt about the existence of soul or spirit; there may be doubt that the mind is anything more than the active brain, but surely there can be no doubt as to the existence of matter itself? It is self-evidently *there* and its properties, though varied and complex, are, at least superficially, indisputable. They are presented directly to our senses.

Yet matter is to some extent puzzling even to common sense, and I am going to invite you to go beyond common sense. I suggest that what at first appear to be two separate questions can be asked about matter, the 'stuff' of the world. First, does matter exist at all? and second, if it does exist, what is its nature, i.e. what are its origins and its structure, and how does it behave?

The common-sense response to the first question is to dismiss it as a pretentious philosophical query. To doubt the existence of matter seems to be nothing more than to make a silly joke; it exemplifies 'a parody of philosophical cautiousness'.[2] Who can doubt the existence of matter? What better evidence can we have than the testimony of our senses? But if we are to appeal to sense experience, we must assess the nature of the evidence it does in fact provide. Our own sense experiences give us sure and certain knowledge of ourselves and of our own existence, but any further claims rest on inference beyond the immediate data of sense. The solipsist refuses to trust such inference and says that she can be confident only of her own existence. However, though solipsism cannot be logically refuted we have no reason to accept it as true.

There is no logical impossibility in the supposition that the whole of life is a dream, in which we ourselves create all the objects that come before us. But although this is not logically impossible,

there is no reason whatever to suppose that it is true;
and it is, in fact, a less simple hypothesis, viewed as a means of
accounting for the facts of our own life than the common-sense
hypothesis that there really are objects independent of us, whose
action causes our sensations.[3]

Russell says that the facts of experience are best explained by
appeal to common-sense belief in an external world of objects
and events, but it should be noted that he refers to that belief as
a hypothesis. Nevertheless, he regards it as a very good hypothesis because it provides a very good explanation of the facts of
experience.

The common-sense belief is also a *basic* hypothesis, in that all
human beings hold it instinctively, without instruction; indeed,
there is evidence that other animals (and not only primates) take
the existence of an external material world for granted. For
example, both wild and domesticated animals will return to
accustomed places for food, horses will hurry on the way to their
stables, many species of bird recognise their mates and offspring,
and many domesticated animals recognise their owners. All such
behaviour shows memory and presupposes a confident belief in
an independent external world of material objects and individuals.

But instinctive beliefs, even widespread instinctive beliefs, are
not necessarily true and certainly not necessarily adequate. Though
we shall dismiss solipsism as self-stultifying it does not follow
that our instinctive and common-sense belief gives us understanding of the nature of matter ('stuff') and the external world. Thus
consideration of the first question 'Does matter exist?' leads to
the second question: 'What is the nature of matter?'

The two questions are related but the second one does not
invite instant dismissal by common sense. On the contrary,
because it presupposes some independent reality impinging on
our senses it invites what we should now call scientific inquiry.
However, as we shall see in chapter 1, both questions already
presuppose sophisticated concepts because the very word 'matter'
involves an abstraction from sense experience. Thus the second
question, 'What is the nature of matter?', has already been
expanded to 'What are its origins and its structure and how does
it behave?' and this invites a further development: 'Are all the
things we perceive composed of a relatively small number of simple
constituents?'

It seems that all early attempts to explain began by appeal to magic and to gods; not surprisingly, the first recorded answers to the double question were in terms of sympathetic magic, myth and primitive religious beliefs. We have some Near Eastern records of myths and priestly incantations from 4,000 years ago. It was probably 1,500 years later that there were first attempts to suggest secular explanations of what could be observed as opposed to invoking divine powers. Progress in this direction was slow, but these attempts were the start of what was to lead to the critical and detailed experiments of physics and chemistry.

Those sciences have shown that matter is far more complex than common sense supposes. Indeed, from the late nineteenth century onwards scientific investigations have led to the development of complex theories as to the ultimate nature of matter which take us far beyond common-sense intuitions. These theories are abstruse and involve appeal to mathematics; accounts of some of the experiments are also complex. It can be difficult for the layman to appreciate both the theories and the experiments, but at least some basic understanding is possible. In this book technical terms are used only when unavoidable and, on first appearance, they are marked with an asterisk to indicate that an explanation is in the Glossary. In many cases, when they are first used there is also an explanation in the text.

Mathematics is necessary to formulate many modern physical theories but here this has been kept to a minimum, to such an extent that apologies are due to those physicists and chemists who could follow a more detailed mathematical exposition. The apologies are necessary because I have had to assume that most readers would find the mathematics confusing and perhaps dispiriting. Even so some may find what is presented too indigestible and for those who wish to eschew all mathematics, the short summaries at the end of each chapter offer conclusions in a simple manner. All chapters have a summary which may be useful for recapitulation as well as providing a convenient (and painless) way of bypassing the technicalities in the last four chapters.

Science was originally part of philosophy, natural philosophy, and modern science requires new concepts in order that we can understand the import of the new discoveries. Part of the aim of this book is to show how, in the study of matter, science and philosophy are still interrelated. The significance of empirical investigations will always depend on assessment and interpretation;

planned experiment and observation, including common-sense observation, must be guided by conjectures that arise from imaginative speculation and analogy. In the course of scientific inquiry we become more and more aware that matter is mysterious and that our beliefs as to its nature depend not only on observation and experiment, but also on our own philosophical theories and metaphysical[4] assumptions. Consideration of the nature of matter is part of philosophical reasoning as well as scientific inquiry, and such consideration shows how philosophy and science are interdependent.

My contention in this book is that inanimate matter and the objects made from such matter are far more mysterious than common sense and even science allow. As we shall see, the mystery arises independently of direct observation and *a fortiori* independently of scientific experimentation. For philosophical reflection as well as scientific theories show that, though the *existence* of an external world is independent of our thoughts, the *nature* of the matter which we believe to constitute that world *is* very dependent on our own interpretations, that is on our theories.

I hope to justify that contention through consideration of various theories (philosophical and scientific) about the nature of matter and of the material world. The discussion will be confined to Western philosophers and scientists, but even within these limits it is not possible to give an exhaustive account of every theory. I have selected those which I believe highlight what I take to be the three concepts that have guided various enquiries into the nature of matter.

All these enquiries have involved attempts to simplify the problem in that they aspire to show that the almost infinite variety of materials observed is to be accounted for by one, or a few, elements (see also p. 10). Starting with early philosophical speculations, discussed in chapter 1, it is possible to trace three types of theory arising from three concepts of matter:

1. Corpuscular theories, based on a picture of sensible materials composed of very small (too small to be seen and therefore invisible) *material* particles. Some theories propose just one kind of particle, and therefore, at the last, one elementary substance, others propose several different kinds of particle and therefore several different elements.

2. Non-material reality theories, based on the view that sensible

objects and materials are a consequence of *immaterial* forces, which themselves constitute an ultimate reality.
3. Idealist theories claiming that sensible objects and materials are but mental constructs.

These three types of theory cannot be sharply distinguished so that in this book we shall see that some of the corpuscular theories discussed in chapters 1, 3, 4, 6 and 7 envisage particles endowed with motion. They may also be related to force and to field theories; others may embody a mathematical idealism. Likewise some of the non-material reality theories discussed in chapters 2 and 8 can be related to corpuscular theories. Lastly, the idealist theories discussed in chapter 5, on phenomenalism, can be related to mathematical theories as to the nature of matter discussed in chapter 8.

The account is more or less chronological, particularly so in the case of corpuscular theories. But because I also wanted to highlight, and to some extent distinguish, the three types of approach, both chapters 2 and 5 take their themes away from the chronological account of corpuscular theories. Within each of these chapters the account is chronological, but I think they needed to be inserted between the corpuscular chapters in order to show the overlap between the three different approaches. The fusion of the three concepts is especially relevant when considering twentieth-century scientific theories as to the nature of matter. It is impossible to appreciate these fully without recourse to mathematics but I have tried to bring out the idealist implications of the equations.

Although I do not advise this, I have suggested that certain technical descriptions in the later chapters can be 'skipped'. However, I do strongly recommend that the many quotations from past writers – Democritus to Dalton, to Bohr and to Heisenberg – be studied. In most cases these are translations, but at least the reader will have a more or less direct account of their ideas and the way they saw their problems.

As indicated above, my treatment does not aspire to be comprehensive, even for the Western world, but I hope it will give some indication of the problem and of the mystery of matter. That mystery may not be resolved, it may not be resolvable, but I hope the fascination of the topic will be revealed and that is perhaps the most important function of philosophy.

The value of philosophy is, in fact, to be sought largely in its uncertainty. The man who has no tincture of philosophy goes through life imprisoned in the prejudices derived from common sense, from the habitual beliefs of his age or his nation, and from convictions which have grown up in his mind without the co-operation or consent of his deliberate reason. To such a man the world tends to be definite, finite, obvious; common objects arouse no questions, and unfamiliar possibilities are contemptuously rejected. . . . [Philosophy] removes the somewhat arrogant dogmatism of those who have never travelled into the region of liberating doubt, and it keeps alive our sense of wonder by showing familiar things in an unfamiliar aspect.[5]

1

Early Theories

It may seem unnecessary to consider early speculations on the nature of matter for those first ideas have been superseded, but it is through studying the first problems and how they were appraised that we can begin to appreciate the difficulties which, even now, are still associated with investigating matter. Moreover, when we read the written records we realise that the theories of nearly two and a half millennia ago were not naive. The very word 'matter', which may seem straightforward, itself presupposes a sophisticated abstraction which we take for granted only because it has become familiar.

We cannot know how our remote ancestors thought about the world around them: there were what must have seemed its permanent features, such as mountains and the sea; then there were the semi-permanent rivers, lakes and trees. But there were also many examples of things showing enormous variety and change in, for example, growth, ageing, death and decay: their own bodies and those of animals and plants. Lastly, there were the heavens showing change but also a remarkable constancy: the movements of the sun, moon and 'stars',[1] the cycle of seasons and local changes in weather. We have evidence that these were observed with great interest from very early times and no doubt sometimes with apprehension.

One characteristic that distinguishes human beings from other animals, and perhaps even from now extinct hominids, was awareness of death. This seems to have developed into a belief that there was life after death, as many sites of prehistoric burials accompanied by burial artefacts have been found. There is evidence that from ancient times it was thought that the spirits of the dead survived in non-material or in tenuous material form (the shades) and that they were able to influence those still alive. There is also plenty of evidence of belief in more powerful living entities (gods) who could influence human life for better and for worse.

1

Thus some sort of distinction between animate and inanimate matter must have evolved, though we must remember that this distinction remained hazy; it was not clear-cut even in the sixteenth century.[2]

In the Preface I said that it is likely that many non-human animals think of the world as having independent reality and this must also have been the case for our hominid ancestors. They would have had concepts of self-subsistent materials and material objects and they would have had concepts of change and of external events. But it is highly unlikely, at least among living creatures today, that any animals apart from humans have an abstract idea of matter. We do not know when the first hazy concepts evolved; there was no word for matter until the fourth century BC (see p. 3 below) but we shall see that the concept itself emerged earlier. As it evolved questions such as 'Could there be some ultimate matter, some basic "stuff", from which all the various materials of that world are composed?' arose. Underlying the complexity and diversity that is observed, we can trace the wish to find simplicity and the conjecture that there might be some homogeneous and fundamental substance.

The word 'substance' can have several meanings but, in the sense taken here, it carries with it the idea that the different materials we observe have properties which are supported by, or which inhere in, some underlying material or substratum. Thus, in this sense, substance is that which lies under (and may support) the overt properties of material things – it is equivalent to 'substrate' or 'substratum'. Today, the theory of a supporting substrate is criticised, but we should appreciate that, even if the theory is now rejected, the word 'substance' like the word 'matter' embodies sophisticated thought about the objects and materials in the world which we come to be aware of through our senses. Those senses give us experiences analogous to those of our ancestors and of other animals, but the significance of those experiences depends on our thoughts. To have concepts of matter and of substance is to abstract from experience; for neither matter nor substance is perceived.

Early ideas about the nature and origin of the world, as shown in the myths and legends of Babylon and Egypt, allude to many deities who could control events. The Ancient Greeks of the sixth century BC were clearly influenced by ideas from the Near East (see note 7) but they developed a new, secular way of thinking.

They were the first to seek explanations which did not rely on appeal to superior powers, that is to gods. There is some appeal to divine participation[3] in many of the early theories, but there are also suggestions of physical mechanisms for the changes observed. They were related to the way that the known craft processes, such as brickmaking, textile dyeing and leather tanning, produced changes. The early Greeks can hardly be said to have developed a systematic method of inquiry into change based on what we would now call 'scientific experimentation', but they did develop ideas about the nature of change. They also developed a language by means of which such changes could be described. This philosophical aspect of inquiry is as much part of scientific method as observation and experimentation and the fact that so many words and expressions that we use today were first developed and used by the Ancient Greeks and Romans shows how much we owe to those first speculations:

> To give some examples: essential and accidental, material and formal, quantitative and qualitative, cause, condition, capacity, potency, necessary, contingent, substance, property, individual, species, theory, hypothesis, purpose, definition, demonstration, deduction, induction, premise, conclusion . . .[4]

Our thoughts and our language are interdependent and the development of a language interacts with the emergence of concepts. The Greek language shows us that the Greeks laid the foundations for expressing those ideas that were an essential basis for our Western science.

However we must bear in mind that some of the words cited above, in particular that translated by 'substance', along with 'matter' and 'element', were not introduced until the fourth century BC[5] and therefore sixth-century philosophers could not have explicitly formulated theories as to the existence of some ultimate primary substance. But though that concept was not explicit we shall see that it was latent, and it developed to become implicit in their theories of matter.

MONISM* AND MYSTICISM – THE MILESIAN PHILOSOPHERS

Early theories on what we would now call the nature of matter and the possibility of an ultimate primary substance would probably be more correctly described as theories as to the origins of material things. First speculations are attributed to three philosophers from Miletus. Theirs was primarily philosophical speculation but their conjectures were to some extent supported by appeal to observation; they show an attempt to relate their theories to materials and changes in the world. The philosophers were: Thales (*c*.624–546 BC), Anaximander (*c*.610–546 BC) and Anaximenes (*c*.585–528 BC). Each proposed a different primary substance from which all materials were formed and the mystical or spiritual aspect of their ideas is apparent in that the primary substance was not to be regarded as inert passive matter but almost as a kind of active spiritual entity with powers of motion and change.

Thales proposed that water was the first material; it was the origin of all the others though this did not necessarily mean that material things were nothing but water:

> it is possible . . . that Thales declared earth *to come from* water (i.e. to be solidified out of it in some way) without therefore thinking that the earth and its contents are somehow water . . .[6]

Aristotle certainly believed that Thales thought water was an ultimate substance, a 'material principle':

> This is the normal interpretation of Thales; but it is important to realize that it rests ultimately on the Aristotelian formulation, and that Aristotle, knowing little about Thales, and that indirectly, would surely have found the mere information that the world originated from water sufficient justification for saying that water was Thales' material principle . . . with the implication that water is a persistent substrate. It must be emphasized . . . that no such development was necessary and that it was not implicit in the near-eastern concepts which were ultimately Thales' archetype. Thales might have held that the world originated from an indefinite expanse of primaeval water . . . without also believing that earth rocks, trees or men are in anyway *made of* water or a form of water. . . . On the other hand Thales

could have made the entirely new inference that water is the continuing, hidden constituent of all things.[7]

Aristotle suggested that Thales may have arrived at his theory by considering that all nutriment is moist and that water is the origin of moist things and hence the basic substance. Thus there was an empirical aspect to his theory, and Kirk et al. think that he earns the title of the first Greek philosopher because he abandoned mythology.[8] But there was still mysticism, a latent appeal to spirits embodied in a certain unphilosophical animism.[9] However, such animism* was a feature of much later Greek philosophy, including that of Aristotle. Thales's mysticism affirmed that everything was full of gods and that the primary substance was not merely the matter of the world but also the spirit which gave the world life. Neither Thales nor the two other Milesian philosophers used the term 'God' in an anthropomorphic sense:

> They wanted a substance which would *explain its own movement*, as in the early days it was still possible to imagine it doing. . . . Thus we find that all of them, while in other respects avoiding the language of religion and completely discarding the anthropomorphism of their time, yet applied the name God or 'the divine' to their primary substance.[10]

Anaximander blended mysticism with observation in proposing an imperishable and everlasting *apeiron* (the boundless) which he said could be described as the initial state of things. He did not specify its properties[11] (later writers called it 'indeterminate matter'), but he said that it could generate 'opposites',[12] such as hot and cold and also motion. Anaximander thought the *apeiron* must be boundless in extent, as well as being eternal, because then, and only then, could it survive destruction and waste. It is not clear whether he had a concept of a fundamental substance:

> While he seems to have gone further than Thales in offering an account of the development of the world, he too, like Thales, may have had no definite view on the question of whether this piece of wood or that piece of bread is the same in substance as the Boundless.[13]

The third Milesian, Anaximenes, proposed air (the Greek *aer*) as the origin of all things and, in contrast to Thales and Anaximander, he did believe that air was the origin of all things. We should bear in mind that '*aer*' meant not only air but also mist and fog.[14] Other materials were formed through a process of rarefaction and condensation of air.[15]

These speculations may strike us as naive and perhaps too fantastic to be treated as serious thoughts about matter, but this is too superficial a judgement. The early philosophers were seeking to explain the nature of the world around them and to understand the significance of what they perceived. Yes, there was mysticism, but any attempt at abstraction can be seen as some form of mysticism since it appeals to something beyond sense experience. The Milesian philosophers show an early, perhaps the first, rational attempt to establish some sort of order amongst the chaotic facts of experience and we can see a steady development of ideas from Thales to Anaximenes and a development of the concepts of matter and of substance.

PARMENIDES

We have seen that the attitude of Thales and Anaximander was ambiguous, but undoubtedly Anaximenes presented a monist theory of matter: air (*aer*) was the one fundamental substance. However, the problem with monism is that just *one* homogeneous kind of matter cannot be made to account for the variety of materials presented to our senses or for the changes we observe. One way of resolving the problem is to deny the reality of change and dismiss the evidence of the senses as did Parmenides (early 5th century BC). He held that knowledge could be acquired only through thought and that thought showed that the world must be a continuous, homogeneous body so that all sensory experiences of change must be delusory.

> do not let habit, born of experience, force you to wander your heedless eye or echoing ear or tongue along this road, but judge by reason . . .[16]

For Parmenides true reality was *non-sensible* and only to be reached by thought; as Guthrie says he 'was the first to exalt the intelligible at the expense of the sensible'.[17] His philosophy is clearly related to that of Plato (see chapter 2) and this was to have a great influence on Greek thought but, at the time, common sense was outraged: the things we see, hear and touch must be real. In effect Parmenides defeated his object because the not unnatural response to his thesis is to reassert a common-sense materialism and then to argue that further inquiry into the nature of matter cannot proceed on monist lines.

FROM MONISM TO PLURALISM* – HERACLITUS

It will be appreciated that Anaximander's theory of the *apeiron* is not purely monistic and a slightly later philosopher, Heraclitus (died after 480 BC), went beyond this and proposed another 'pluralistic monism'. He was not prepared to dismiss sensory evidence of variety and change; rather, he affirmed that there was permanence *in* change; change was the fundamental basis and explanation of what happened in the material world.

For Heraclitus fire was the primary substance, the archetypal form of matter,[18] because fire was the principle of change. Fire, he said, shows permanence and change: the flame appears to be a permanent entity, but it is continually changing as it burns. Thus fire and flame exemplify the idea of a single world composed of changing matter. The paradox is made plainer by Heraclitus's comparison of flame with a river: just as flame seems permanent but is continually changing, so a river seems permanent though the flowing water is continually changing:

> The great 'discovery' on which Heraclitus so prides himself was the identity of the One and the Many: 'out of all one and out of one all' . . . it is not sufficient, Heraclitus says . . . to suppose some external One, from which the substance of the world is supplied; rather it is the world itself which is One, and its unity is in fact constituted by the very multiplicity of things, which at first sight seems to destroy it. In the world . . . we see perpetual flux and constant change: 'all things pass away and nothing abides' . . . Yet behind this unceasing flow and constant alteration there is permanence; the ever-shifting things of experience are but parts of an abiding whole.[19]

Pluralism does not necessarily entail a theory of a *limited* number of primary simple substances. Anaxagoras (*c*.500–428 BC), living a little later than Heraclitus, also wished to explain change in terms of underlying permanence. Familiar examples of change were extraction of metallic iron from iron ore[20] and the production of body tissues from food. He suggested that these were due to the rearrangement of their constituent 'seeds'. The seeds were infinite in number and there were also an infinite number of different kinds of seeds. This was a thorough-going pluralism. For Anaxagoras bones were made up of seeds of bone tissue, gold from seeds of gold, water from seeds (in this case tiny droplets) of water, and so on. But though each material was largely composed of its characteristic seeds, seeds of every other kind of material were also present in it, for, he said, there was a portion of everything in everything.

> But before these things were separated off, while all things were together, there was not even any colour plain; for the mixture of all things prevented it, of the moist and the dry, the hot and the cold, the bright and the dark, since there was much earth in the mixture and seeds countless in number and in no respect like one another. And since this is so, we must suppose that all things were in the whole.[21]

Thus every material had a potentiality for changing into every other material. For example, though seeds of body tissue cannot be seen in bread they must be there; it is just they are too small to be detected.

By appeal to his seed theory, Anaxagoras wanted to show that no really new material could be produced, but there are obvious difficulties inherent in his account. For example, what happens to all the bread seeds when bone is formed? At least some of them must go elsewhere for otherwise the bone would *be* bread. Perhaps there is a problem of principle in that if there are indeed pure bread seeds and pure bone seeds then, within each seed, there is *not* a portion of everything in everything. On the other hand, it has been suggested that Anaxagoras may have thought that at the microscopic level of individual seeds, each was homogeneous.[22]

ULTIMATE SUBSTANCES – ELEMENTS*

The more fruitful proposal that there was a relatively small number of basic substances provided a better explanation of change. Empedocles (*c*.484–424)

> expresses more clearly than any earlier writer the idea of sub-stances that are both original and simple. True he does not use what became the technical term for element in Greek; *stoicheion* which is not introduced until Plato, but he refers to earth, water, air and fire as *rhizomata*, 'roots', in a well-defined sense.[23]

These, which we now call 'elements', were indestructible and unchangeable. They were constituents of all materials so that changes observed were due to rearrangement and to different proportions of the elements. Although none of Empedocles's elements is now regarded as a simple material, his conception of an elementary set of substances still survives and is still of primary importance. He made a basic contribution to chemical theory.

In choosing his four elements Empedocles had probably been influenced by earlier philosophers: Thales's basic substance was water, Anaximenes's was air and Heraclitus's was fire. But Empedocles's elements, though simple substances, were not chemically pure in the modern sense and they could perhaps better be regarded as embodying the states of matter: earth (solid), water (liquid), air (gas) and fire (a still more rarefied state). Empedocles's four-element theory, as developed by Aristotle, was to dominate ideas into the seventeenth century AD. For example, when dealing with the origin of metals Agricola (1494–1555) writes:

> I must explain what it really is from which metals are pro-duced. The best proof that there is water in their materials is the fact that they flow when melted, whereas they are again solidified by the cold of air or water. This, however, must be understood in the sense that there is more water in them and less 'earth'; for it is not simply water that is their substance but water mixed with 'earth'. And such a proportion of 'earth' is in the mixture as may obscure the transparency of the water, but not remove the brilliance which is frequently in unpolished things. Again, the purer the mixture, the more precious the metal which is made from it, and the greater its resistance to

fire. But what proportion of 'earth' is in each liquid from which a metal is made no mortal can ever ascertain.[24]

STRUCTURE – PYTHAGORAS

Elements are simple basic substances, but another aspect of the problem of the nature of matter is the question of the structure of the simple substances. Again, early speculations were primarily mystical and metaphysical*: were the elements composed of indivisible units and, if so, what were these units like? According to Aristotle, Pythagoras (c.571–497 BC) had suggested that the cosmos and the visible world could be explained in terms of number:

> the so-called Pythagoreans, who were the first to engage in mathematics, advanced this study, and being trained in it they thought that its principles were the principles of all things. But of these principles numbers are by nature the first, and in numbers they seemed to see many resemblances to the things that are and come to be – more than in fire and earth and water . . . Therefore, since all other things seemed in their whole nature to be modelled on numbers, and numbers seemed to be the first things in the whole of nature, they supposed the elements of numbers to be the elements of all things . . .[25]

Pythagoras thought of numbers as shapes as they appear on dice and we still speak of 'squares' and 'cubes', terms which we owe to him. He classified numbers as oblong, triangular, pyramidal, and so on. This was because the numbers were represented by the same symbol, or by pebbles. It is clear that to appreciate numbers of any size the pebbles would have to be arranged in a pattern. The early Pythagoreans did suggest that

> 'things were numbers'. To demonstrate it they said: 'Look! 1 is a point (.), 2 a line (—), 3 surface (□), and 4 a solid (▱). Thus you have solid bodies generated from numbers' We may call this an unwarrantable and indeed incomprehensible leap from the abstract intellectual conceptions of mathematics to the solid realities of nature. . . . [But] Numbers in fact, like everything else – whether objects or what we should distinguish

from objects as mere conventional symbols, words or names – are endowed with magical properties and affinities of their own.[26]

Pythagoras did not propose a theory of matter itself, rather of the arrangement of number, but, stripped of mystical overtones, his ideas can be interpreted as a proposal that physical natures are grounded in different geometric structures and this influenced atomic and corpuscular theories, which were to develop later.

The importance of these speculations has been acknowledged by the twentieth-century scientist Heisenberg, who contributed to changing, fundamentally, our notions of the nature of matter (see chapter 7). What he writes summarises some of the points made above:

The first physical phenomenon to attract the attention of Greek systematic thought was that of 'substance', the 'lasting' element of the mutations of all phenomena. In the thesis of Thales, that water is the fundamental substance of which the world 'consists', we can see the formulation of the concept of 'matter'. . . . None of the words 'fundamental substance', 'water' or 'consist' had a concisely defined field of application or an unambiguous meaning, . . . Succeeding research defined the term 'fundamental substance' somewhat more concisely [sic]. First it acquired the characteristic of uniformity and indestructibility. This formulation resulted in a complication, for in order to make intelligible the changing phenomena of the world, one had either to assume several fundamental substances . . . or to separate altogether the concept of 'lasting' from common experience. Parmenides's idea of 'being' was an attempt in the latter direction. Empedocles regarded earth, fire, air and water as the four 'basic roots' of all things. He considered them 'uncreated, indestructible, homogeneous, immutable but at the same time divisible'. . . . Anaxagoras postulated an infinite number of elements. . . . This work prepared the ground for an explanation of the qualitative variety of the external world in terms of variations of quantity and changes in the proportions of the mixture. . . .[27]

ATOMS AND THE VOID: A NEW MONISM – LEUCIPPUS AND DEMOCRITUS

If only one simple homogeneous substance is postulated then, as we have seen, it seems impossible to account for change and variety and we must dismiss the evidence of our senses as illusory. The difficulty may be avoided by suggesting that the basic substance is composed of small particles, separated by empty space (the void) as then, since the particles can move and change positions there is scope for variety as well as change. There is, then, an absolute distinction between the concepts of space and matter. Atomic theories also entail that the particles be indivisible so that they can be regarded as basic blocks from which all materials are to be formed. Such a theory was suggested by Leucippus in the fifth century BC, though none of his writings survive. It is alleged that he thought that though atoms were too small to be seen, they must be of different sizes and shapes in order to account for different materials. But perhaps Leucippus's most innovative idea was to affirm that empty space (the void) was as much an existent as matter itself. He suggested that space was 'porous' and its reality depended on its being a vehicle for accommodating atoms.

Leucippus's ideas were developed by Democritus (c.460–371 BC) who used the atomic theory to explain various natural phenomena such as thunder, lightning and earthquakes. These explanations were no more than conjectures but they were attempts to give a mechanical explanation of events. For Democritus, change was due to the constant movement and rearrangement of atoms; like Leucippus, he thought these were eternally in motion and though they were all of the same basic substance, they differed in size and shape:

> these atoms move in the infinite void, separate one from the other and differing in shapes, sizes, position and arrangement; overtaking each other they collide, and some are shaken away in any chance direction, while others, becoming intertwined one with another according to the congruity of their shapes, sizes, positions and arrangements, stay together and so effect the coming into being of compound bodies.[28]

Democritus gave no explanation for this motion; atoms obeyed a law of their own being, and it was no more sensible to ask for the cause of their motion than to ask why matter existed in the form of atoms. The metaphysical principle underlying his theory was that all change was due to the movement of the immutable primordial atoms:

> Just as tragedy and comedy can be written using the same letters, so, many varied events in this world can be realized by the same atoms as long as they take up different positions and describe different movements.[29]

Not only did he explain natural events, but he also used the theory to explain properties of matter such as density and changes of state*; he even drew a distinction between what we should now call primary qualities, which were inherent in the material and directly related to the size and shape of its atoms, and secondary qualities, which were powers to produce appropriate sensations by acting on our sense organs.

> Sweet exists by convention, bitter by convention, colour by convention; atoms and Void (alone) exist in reality...[30]

> What we feel as hard has its atoms closely packed. Soft things are made of atoms wider apart, they contain more empty space and so are capable of compression and offer less resistance to the touch. The other senses are explained on the same lines. In taste, sweet things are made of smooth atoms, whereas harsh or bitter flavours are caused by hooked or sharp-pointed atoms which tear their way into the body making minute excoriations on the tongue.[31]

Since atoms themselves do not possess secondary qualities (colour or taste, for example) these qualities lose the immediacy they have in our ordinary experience. Thus one consequence of any atomic theory is that those secondary qualities have to be represented in abstract form as certain geometrical and dynamic configurations of the ultimate particles. But this abstraction means that the qualities themselves cannot be analysed by science:

> The desire fulfilled in atomic theory, to depict perceptible qualities of things like colour and hardness, by means of reduction to geometrical configurations... enforces the sacrifice of

ascertaining the true nature of these qualities by means of science. Thus it can be easily understood why the poets for example always looked upon the atomic concept with horror.[32]

But there could also be a latent and often an explicit materialism in these atomic theories. The theory of atoms and the void was ardently elaborated by Lucretius (97–54 BC) who, by appealing to mechanical explanations based on atoms whirling in the void, made it a basis for a materialist philosophy. There was no independent place for soul or spirit. Not surprisingly, the early Fathers of the Christian Church attacked such notions vigorously, and atomic theories of matter along with the distinction between primary and secondary qualities were not reconsidered until the seventeenth century. (See also note 36.)

MATTER AND FORM – ARISTOTLE

Well before Christian times Democritus's atomism had been criticised by Aristotle (384–322 BC) on secular grounds. Since his influence was so great this was another important reason for the rejection of atomism. It has been suggested that Aristotle did entertain some theory of smallest parts,[33] but he did not develop the idea and it is difficult to see how he could have accepted any form of atomism as conceived by Democritus since he was convinced that there could be no empty space and hence atoms could not be moving through the void. Two of Aristotle's reasons for denying the existence of a vacuum emerged as a consequence of his mechanics. He said that the speed of any falling body was inversely proportional to the resistance of the medium (air, water, for example) through which it fell; therefore, in the void a body would fall with infinite speed and this was clearly impossible. So a vacuum was impossible. Also, if a void existed, a moving body subject to no external forces would move forever; since Aristotle and his contemporaries had no idea of inertial motion they regarded perpetual movement as impossible and hence this was yet another reason for rejecting a theory postulating empty space.

   Aristotle's elaboration of the four-element theory was part of his conception of substance in terms of matter and form and of his account of change in terms of potentiality and actuality. He made no attempt to relate properties to structure.

For Aristotle there were several significations of the word 'substance'; here we are primarily concerned with the meaning outlined above, that is of substance as a substratum in which properties are carried or in which they inhere. Aristotle further developed the concept to include the notion of substance as a centre for change. In that sense a substance must have an independent existence and is to be distinguished from qualities or properties that cannot exist independently. Qualities must be parasitic on something else: for example, colour (say, yellowness) or texture (say, smoothness) must inhere in or be a property *of* something, that is of a substance. But can that supporting substratum, the basic substance, itself be independent? Can it exist without qualities? Even if it could, we could not know it, for it would have no distinguishing feature and clearly it could not be identified.[34]

To deal with the problem of substance and qualities Aristotle developed the concepts of matter and form and of potentiality and actuality. Form was the knowable aspect of anything, what we could be aware of and what could be described, defined and classified. It was pure actuality, a determining principle of being. Pure form could exist on its own for it was believed that non-material spirits and gods were pure form.[35] Of course, all materials and all material objects also possessed form: but for them the form inhered in a primary matter, *materia prima*, which could be regarded as an unknowable and structureless residue. It was pure potentiality.[36]

Unlike form, primary matter could not exist independently;[37] it had to have some qualities, some form. But different forms might be imposed on matter and, as one form superseded another in the process of change, that form was the actuality of the previous potency. Thus matter and form, regarded as participants in the process of change, become potentiality and actuality – potency and act. Matter and form are correlative notions:

A silver bowl may be analyzed into its matter – silver – and its form – the structure given to the silver by the craftsman who made the bowl. [At this level 'form' signifies shape.] But a piece of unworked silver provides the same distinction. It has an observable character, different ... from ... gold or copper. This is its form. And its matter ... consists of the elements out of which the silver must ultimately be composed – fire, earth, air, and water. The *proportions of the elements present in the silver are, of course, part of the form.*[38]

The elements themselves were composed of matter, in this case *materia prima*, and their form was a blend of two of the four primary qualities: hot, cold, wet and dry. Fire was a blend of hot and dry, earth cold and dry, air hot and wet, and water cold and wet. These elements were the basic substances but they were not immutable;[39] they could be changed into each other. Moreover, Aristotle related them not so much to observed samples of earth, water, air and fire, but to the states of matter: solid, liquid, gas and flame. We have seen how this was accepted by Agricola (see above note 24).

Aristotle made many references to the everyday materials of his time but, for him, changes brought about in craft processes (what we regard as chemical changes) would have been changes of form. His account of such changes can therefore allow for a new material which is not just the sum of its parts, for there can be a new form imposed even though the new product is still composed of the four elements. Thus his notion of chemical change was more sophisticated than that of Democritus who explained change in terms of new configurations of atoms. For Aristotle the change was an internal change, something more than a rearrangement.

The idea of spontaneous internal change was especially developed in relation to changes in living things, for example the acorn becoming an oak, the child becoming an adult. Here the potentiality was inherent. It is possible but unlikely that Aristotle envisaged maturation or decay in terms of chemical actions analogous to the smelting of metals or the baking of clay, though both types of change illustrated potentiality becoming actuality.

The Ancient Greeks were possibly the first people to become aware of the problems of matter and to formulate them so that they could be tackled. Starting with the Milesians we can see how their ideas and theories developed, how problems arose and how they were treated. Many of those problems are still with us today. What *are* the basic constituents of matter, how are they arranged and how much do our descriptions depend on our own ideas? To what extent can we trust the evidence of our senses? Can they lead us to some ultimate reality?

SUMMARY

The concepts of matter and of substance are sophisticated abstractions from sense experience and the earliest recorded speculations about the nature of matter are those of three Milesian philosophers: Thales (624–546 BC) and Anaximander (610–546 BC), who both suggested monist theories, and Anaximenes (585–528 BC), who proposed a more abstract 'indeterminate matter'.

A little later Parmenides (fl. 501–492 BC) thought that ultimate reality was non-sensible and that change was illusory, whereas Heraclitus (fl. 504 BC) suggested fire as an unchanging principle of change.

A theory of there being a relatively small number of basic substances, the elements, was proposed by Empedocles (484–424 BC). Pythagoras (571–497 BC) proposed that different physical natures were grounded in different geometric structures.

The first theories of atoms and the void were those of Leucippus (fifth century BC) and Democritus (c.460–371 BC) These were not acceptable to the Christian Church but were revived in the seventeenth century. Leucippus and Democritus sought support for their atomic theories from direct observation. They did not offer purely philosophical speculations. They also first introduced the notions of primary and secondary qualities.

Aristotle's account of the correlative notions of matter and form (potentiality and actuality) came to dominate later theories of matter. Aristotle also developed Empedocles's theory of four elements.

These early theories have been superseded but, as far as we know, the Greeks were the first to formulate the problem of matter and many of the questions they asked are still with us today.

2

Non-material Reality

The four philosophers considered in this chapter, Plato, Leibniz, Boscovich and Kant, had a common belief that reality was not directly accessible to the senses. However, they did not think that the materials and objects perceived were necessarily illusory; rather, they held that their existence depended on something that transcended the material world and had to be known, if it could be known at all, through reason, not through sense perception. For Plato, material entities were but (metaphorical) shadows of immaterial Forms; for Leibniz they existed as a consequence of a confused understanding (perception) of spiritual monads; for Boscovich they were composed of indivisible and non-extended points of force; for Kant they were our constructs (phenomena*), and ultimate (noumenal*) reality was not accessible even to reason. These views may seem to offer a perverse challenge to common sense but they have profoundly influenced scientific as well as philosophical theories as to the nature of matter.

PLATO

Plato (427–347 BC) agreed with Heraclitus that there was perpetual change in the world; there was no stability. He stressed that *all* the objects we perceive, not just things like shadows and rainbows but tangible entities such as horses, mountains or people, lacked permanent properties and had no permanent existence.[1] Hence propositions about material things could not be eternally and therefore *necessarily* true, for they could be true at one time and false at another time. In addition the senses, although not always delusory, could be unreliable. Therefore, for two reasons, any proposition about material things might be mistaken.

Plato insisted that objects of knowledge had to be permanent and unchanging because then true statements about them would indeed be eternally and necessarily true. Like Parmenides he held that knowledge came through thought and necessary truths* were known to *be* true because they are apprehended as such by the mind.[2] Their necessary nature did not depend on the evidence of sense perception. Examples of necessary truths are arithmetical propositions such as '2 + 2 = 4'; and so for Plato these were proper objects of knowledge. But there were other objects of knowledge: those which constituted the stable and permanent reality underlying the material entities we perceive with our senses.

Plato invites us to contrast two types of things: those that can be perceived and those that can be conceived but *not* sensed. These latter are the Forms. The Forms (which are sometimes called *Ideas* and sometimes *Universals*) are like mathematical entities in not being accessible to sense perception. They can be regarded as metaphorically embodying the non-material essences of the various types of material, or thing or quality in the world. But for Plato they were more than this because they were the ultimate reality. Thus there is a Form of gold which *is* gold, a Form of horses which *is* the horse, a Form of beauty which *is* beauty, etc. Particular pieces of gold, individual horses and beautiful objects or persons have their attributes by virtue of their participation in the appropriate Form. Moreover the very existence of particular entities – all particular pieces of gold, all particular horses, the beauty of all beautiful objects – is dependent on there being an appropriate Form.

However, Forms must not be taken to be concepts or ideas in the sense that we use the words today, for concepts and ideas are dependent on our minds and are produced by our mental activity. By contrast Plato's Forms were held to exist quite independently of human beings and their thoughts. Nor must they be confused with Aristotle's notion of form. For Plato the Forms were *the* permanent objective reality to which, if we applied our mental powers, our concepts might correspond. Hence the terms 'Form' (which implies either shape or Aristotelian form) and 'Idea' can be misleading and it is less confusing to use the word 'Universal', although that is not how Plato describes them.

For Plato, Universals were immaterial and completely independent entities which existed in an immaterial world of Universals. This world and the Universals which made it were the ultimate and objective reality. Plato held that whatever reality was ascribed to the world that we are aware of through sense perception, the things we see, hear, touch, etc., were due to the fact that the objects of sense perception shared, though to a limited and imperfect extent, in the nature of Universals. Plato thought he could not demonstrate the existence of Universals by rational deductive argument and that, at bottom, his account had to be accepted on the basis of an intuitive belief that ultimate reality was immaterial and not accessible to sense perception. But we should not, on that account, dismiss the Platonic conception of Universals and of knowledge. For *any* account that we give as to the nature of the world must ultimately rest on intuitive beliefs which cannot be logically justified.

Plato had to rely on metaphor and simile, and his allegory of the cave is the most famous of these:

> Imagine mankind as dwelling in an underground cave with a long entrance open to the light across the whole width of the cave; in this they have been since childhood, with necks and legs fettered, so they have to stay where they are. They cannot move their heads round because of the fetters, and they can only look forward, but light comes to them from fire burning behind them higher up at a distance. Between the fire and the prisoners is a road above their level, and along it imagine a low wall has been built, as puppet showmen have screens in front of their people over which they work their puppets.
>
> . . .
>
> See, then, bearers carrying along this wall, statues of men and other living things, made of stone or wood and all kinds of stuff, some of the bearers speaking and some silent, as you might expect.
>
> . . .
>
> Suppose the prisoners were able to talk together, don't you think that when they named shadows which they saw passing they would believe they were naming things?
>
> . . .
>
> Now consider . . . what their release would be like, and their cure from these fetters and their folly . . . One might be released and compelled suddenly to stand up and turn his neck round, and to walk towards the firelight; all this would hurt him, and he would be too much dazzled to see distinctly those things whose shadows he had seen before.
>
> . . .

Suppose . . . that someone should drag him thence by force, up the rough ascent, the steep way up, and never stop until he could drag him out into the light of the sun, would he not be distressed and furious at being dragged; and when he came into the light, the brilliance would fill his eyes and he would not be able to see even one of the things now called real?
. . .

[He would at last become accustomed to the brilliance and would see clearly]
. . .

Let him be reminded of his first habitation, and what was wisdom in that place, and of his fellow-prisoners there; don't you think he would bless himself for the change, and pity them?
. . .

The world of our sight is like the habitation in the prison, the firelight there to the sunlight here, the ascent and the view of the upper world is the rising of the soul into the world of mind;[3]

In this allegory Plato showed that he was well aware that our instinctive belief is that the external world which we 'know' through sense perception is the real world. Through metaphor he hoped to convince others that this belief was mistaken. If we rely on sense perception, we remain like the prisoners in Plato's cave and are aware only of shadows (the material things we sense), which we take to be reality. But if we use our understanding, then the fetters of sense perception may be broken.

It must again be stressed that Plato did not think that matter did not exist. He had compared material objects to shadows in the cave, and the shadows certainly existed. In the *Timaeus* it seemed he regarded the study of nature (what he called 'the world of becoming') as recreation,[4] though there was an ethical motive in that he wanted to stress the role of an intelligent guiding agency in the universe.[5] He adopted Empedocles's four-element theory and developed this into a geometrical analysis, each element being identified with one of the regular solids:[6] fire with the tetrahedron, air with the octahedron, water with the icosahedron and earth with the cube.[7]

The influence of the Pythagoreans on Plato is pervasive.[8] This has been noted by Heisenberg,[9] who pointed out that Plato had referred to Pythagorean investigations into the harmonics of vibrating strings. There was a difference though in that the Pythagoreans were primarily interested in the numerical relations, whereas for Plato it was the melody itself, as participating in the Universal. For him the study of mathematics was but a prelude to understanding the nature of things and the Universal underlying the particulars.

Plato was also influenced by the atomists:

> The differences between Plato's theory and that of the atomists are also instructive. He owed to them the idea that the varieties of sensible objects can be reduced to differences in the shapes and sizes of particles that are themselves homogeneous in substance. But while they thought of the basic particles of matter as solid, Plato suggested that the primary solids are in turn composed of plane surfaces[10] ... secondly while the atomists postulated a void, Plato denies this and evidently saw that in a plenum motion is possible provided it is (i) instantaneous and (ii) cyclical. [He also believed in the possibility of transmutation]. ... he makes a number of specific suggestions about for example how water may be decomposed into fire and air. Thus the icosahedron of water may become two octahedra of air and one tetrahedron of fire ...[11]

Guthrie says that Plato must have been aware of Democritus's views even though he would have been antipathetic to the implied irrational mechanism.[12]

As Heisenberg pointed out:

> basically Plato thought only of a single entity which happened to appear in different shapes, Nature's variety was the result of the diversity of mathematical structures.[13]

Plato's belief in the mathematical construction of the units making up particular entities still left them as pale shadows of the Universals but even so they were, in his view, the best possible:

> While he often contrasts the world of becoming unfavourably with the Forms, he nevertheless repeatedly asserts that this is the best possible created world. It is the fairest of things that come to be, its maker is good, it is fashioned after the most perfect model and it is as like that model as possible.[14]

We do not know how original these ideas were, that is, the extent to which Plato was relying on his predecessors and his contemporaries,[15] but in one respect at least his theory of the material world influenced a much later philosopher, Leibniz, for he too thought the world which had been created was the best possible world.

There are criticisms of Plato's view of knowledge and his view that only necessary truths are proper objects of knowledge. There have been criticisms of his realist account of Universals and their relation to the particular entities we sense. Plato himself was vague as to this relationship and his pupil Aristotle proposed a different account of Universals. Yet it remains true that Plato did raise problems concerned with our knowledge of the material world and the problem of the relation between sense perception and reasoning which are still with us today.

He expounded a metaphysical view of the ultimate nature of the world and a thesis that knowledge comes solely through ratiocination, which at first reading seems utterly alien to scientific investigation. He appears to lead us into a cul-de-sac analogous to that of Parmenides. But his discussion of the possibility of some permanent reality is not self-defeating and sterile. For Plato, the Universals supported and gave shadowy reality to what could be perceived and we may say that our science, by striving to explain and account for what we observe, also postulates something beyond sense perception and the inferences of common sense. We too seek a reality underlying appearances and that reality is *not* directly accessible to the senses. So though we would not today accept an explanation of the properties of a material we perceive, say gold or bread, by referring to a corresponding Platonic Universal, any explanation we do entertain will involve theoretical constructs.

LEIBNIZ

As was pointed out in chapter 1, 'matter' and 'substance', when used as technical terms, are theoretical constructs. Gottfried Wilhelm Freiherr von Leibniz (1646–1716) lived more than 2,000 years after Plato and though he did not propose a Platonic theory of Universals, he agreed with Plato that knowledge of ultimate reality could not be established by sense perception. Leibniz argued that:

there was no way of determining, by a mere consideration of the motion of a set of bodies moving relative to one another, which of the motions were real ones.[16]

This was a reason

> for concluding that there is something not completely real – something phenomenal – in motion, and for deciding that the nature of substance must be found in something other than body (since body gets all its positive attributes through motion).[17]

But originally he had agreed with most of his forward-looking contemporaries, who were influenced by Descartes:

> for whom extension, figure, and motion were the outstanding examples of what is clear and distinct, not only making rational investigation possible in science but also justifying the claim that what science was investigating was the nature of reality itself.[18]

At that time Leibniz had been influenced by the writings of the atomists. But further reflection led to an additional reason for him to be critical of materialist theories since he came to appreciate that it was impossible to show whether matter was continuous or atomic; both views involved contradictions. He finally concluded that there must be something more to matter than physical extension:

> At first, when I had freed myself from the yoke of Aristotle, I had believed in the void and atoms, for it is this which best satisfies the imagination. But, returning to this view after much meditation, I perceived that it is impossible to find *the principles of true unity* in matter alone, or in what is merely passive, since everything in it is but a collection or accumulation of parts *ad infinitum*. Now a multiplicity can be real only if it is made up of *true unities* which come from elsewhere and are altogether different from mathematical points, which are nothing but extremities of the extended and modifications out of which it is certain that nothing *continuous* could be compounded. Therfore to find these *real unities*, I was constrained to have recourse to what might be called a *real and animated point* or to

an atom of substance which must embrace some element of form or of activity in order to make a complete being. . . . I found then that their nature consists of force and that from this there follows something analogous to feeling and to appetite; and that therefore it was necessary to form a conception of them resembling our ordinary notion of *souls*. [Leibniz was to modify this; see note 24 below]. . . . Aristotle calls them *first entelechies*; I call them, more intelligibly perhaps, *primitive forces*, which contain not only the *act*, or the fulfilment of possibility, but also an original *activity*.[19]

Though Leibniz came to think of

extension figure and motion as something imaginary and [he] held that the basic laws of motion [could not] be derived from a study of their nature. Nevertheless he continued to hold that, given the basic laws derived from other sources, extension, figure and motion could produce a means for explaining and predicting the course of phenomena.[20]

Thus, though the contemporary view was to treat matter as inert, for Leibniz this was true only when dealing with phenomena. Ultimately, the essence of substance was activity. This has been explicated by Bertrand Russell:

Matter as such is extended; extension is essentially plurality; therefore the elements of what is extended cannot themselves be extended. A simple substance cannot be extended since all extension is composite. . . . Hence the constituents of matter are not material if what is material must be extended. But the constituents cannot be mathematical points, since these are purely abstract, are not existents, and do not compose extension. The constituents of what appears as matter, therefore, are unextended and are not mathematical points. They must be substances, endowed with activity, and differing *inter se* . . . Hence there remains nothing, among the objects of experience, which these substances can be, except something analogous to souls. . . . Bodies as such, *i.e.* as extended are phenomena; but they are *phenomena bene fundata*, because they are the appearances of collections of real substances. The nature of these is force,[21] and they are indivisible like our minds.[22]

For Leibniz all extended bodies were aggregates of parts and therefore could not represent a single, simple entity – a unity. But there could be unity in a centre of activity, that is in an entelechy or mind. (Leibniz regarded minds as unities.[23]) Like many of his contemporaries he had been fascinated by animaliculae* revealed by the microscope and it may well be that his belief in ultimate points of energy (and of life) had been influenced by the discoveries of Swammerdam, Malpighi and Leeuwenhoek in the seventeenth century.[24]

Therefore, though ultimate reality was immaterial, Leibniz gave tangible and visible matter a phenomenal reality (*phenomena bene fundata* see quotation from Russell above). This was, after all, the subject of his mechanics and of other sciences, such as biology and the creatures observed by the microscopists. But, as we have seen, he held that the ultimate basis of matter was an infinite number of *non-material* souls which he called monads. Each monad was a unity, but it had both a passive and an active aspect; the passive aspect he sometimes called *materia prima* and *in the immaterial monad materia prima* was equivalent to passivity, or confusedness of perception (see below). The active aspect was force (entelechy) which could also be regarded as mind or soul. For Leibniz, force (entelechy) was related to *materia prima* as Aristotelian form was related to matter, and therefore *materia prima* was an abstraction, just as matter was for Aristotle (see p. 15). The passivity and activity were entirely internal for no monad interacted with any other monad; they simply 'mirrored' or 'perceived' the other monads and it was confusion of their perceptions which gave rise to their passivity.

The aggregates of monads made up what Leibniz called *materia secunda*. For any aggregate there was a dominant monad which had an especially active entelechy or soul, whereas the entelechy of other monads was relatively passive. 'There is no particle of matter, however small, that is not composed of organisms, each with its dominant monad.'[25] Thus, for Leibniz, all matter was, in a sense, animate.[26]

We need to bear in mind that Leibniz used the phrases *materia prima* and *materia secunda* in two different senses: a metaphysical and a phenomenal sense.[27] As we have seen, in the metaphysical sense the former was an element in the nature of every monad and the latter was an aggregate of monads with a dominant monad. In the phenomenal sense *materia prima* was an abstraction presupposed by extension and *materia secunda* was a corporeal substance, the seat of physical forces and the subject of study for physics and, in particular, of dynamics. It appears *to us* as a body because *we* have confused perception. Broad writes:

> It is important to notice that, on Leibniz's theory, the appearance of corporeal substance depends on a double confusion. (a) A set of monads will not be perceived by any mind as a corporeal substance unless *they* are all extremely confused. (b) A mind will not perceive any set of confused monads as a corporeal substance unless *it* is itself somewhat confused. If the percipient were free from confusion, he would perceive such a group correctly, viz, as an infinite collection of very confused monads. . . . No doubt, e.g., God perceives those sets of monads which we misperceive as bodies. . . . He perceives then correctly as groups of confused monads.[28]

Thus all material bodies (well-founded phenomena) were not ultimately real but only some kind of 'shadow' of the metaphysical monads.[29] We may note another influence of Plato. Leibniz himself wrote:

> Extension and motion and bodies themselves, in so far as they consist of these alone, are not substances but true appearances, like rainbows and mock suns, for shapes do not exist objectively, and bodies, if they are considered as extension alone, are not one substance but several.
>
> For the substance of bodies something without extension is required.[30]

In dismissing the equivalence of 'body' with extension, Leibniz is alluding to Descartes' dualism which we shall consider in the next chapter, In his *Monadology* Leibniz wrote:

> We may give the name *entelechies* to all created simple substances or monads. For they have in themselves a certain perfection, there is a self-sufficiency in them which makes them the sources of their internal actions.
>
> . . . .
>
> If we wish to give the name 'soul' to everything which has *perceptions* and *appetites* in the general sense . . . all created simple

substances or monads might be called souls; but as feeling is
something more than simple perception, I agree that the general
name – monad or entelechy – should be enough for simple
substances which have no more than that, and that only those
should be called souls, whose perception is more distinct and
is accompanied by memory.[31]

We may feel that Leibniz shows a certain tendency to have his cake
and eat it in postulating immaterial ultimate reality and yet also
dealing with material bodies as well-founded phenomena. Other
writers, for example Bertrand Russell,[32] have criticised his account
of monads and their 'points of view' since it embodies assumptions
as to the objective reality of space which Leibniz elsewhere denies.
But it is interesting to consider his account of matter because the
concept of entelechy, and of non-material monads, may be related
to our present conceptions of ultimate particles as carriers of energy.
At present we constantly 'discover' new sub-atomic particles which
seem close to being units of energy. Leibniz's assertion that the
basis of matter must be non-material, energy rather than mass,
might seem not too far removed from our present theories. Such
a connection is intriguing but it is very tenuous for, as L.J. Russell
says, 'It does not seem there is any way of making out Leibniz's
views in detail'.[33] (See also Recapitulation in chapter 10.)

BOSCOVICH

In developing his theory that the ultimate bases of matter were
points of force, Roger Joseph Boscovich (1711–87) comes much
closer to twentieth-century science. Like Leibniz he thought that
the simple components of matter were non-extended immaterial
discrete points and he explicitly acknowledged his debt to Leibniz.
But though Boscovich was a Jesuit priest he did not elaborate a
separate immaterial metaphysical world; nor did he relate energy
to soul. Boscovich also acknowledged a debt to Newton in that
he held that at a certain distance there would be an attractive
force between the ultimate points. In other words, and in contrast
to Leibniz, interaction was possible. But Boscovich stressed that
there were other important differences between his theory of
elementary parts and that of Leibniz, and also between his theory
and the Newtonian theory of force.

Apart from his metaphysics, Boscovich differed from Leibniz in that he postulated points which were identical and also separated. He argued that chemical analysis was beginning to show that the vast variety of different materials were composed of a relatively small number of elements and he hoped that further analysis would show that there was an even smaller number of ultimate constituents – perhaps only one:

> But my Theory differs in a marked degree from that of Leibniz. For one thing because it does not admit the continuous extension that arises from the idea of consecutive, non-extended points touching one another; here, the difficulty raised in times gone by ... & never really or satisfactorily answered ... with regard to compenetration of all kinds with non-extended consecutive points, still holds against the system of Leibniz. For another thing, it admits homogeneity amongst the elements, all distinction between masses depending on relative position only, & different combinations of the elements; for this homogeneity amonst the elements, & and the reason for the difference amongst the masses, Nature herself provides us with the analogy. Chemical operations especially do so; for, since the result of the analysis of compound substances leads to classes of elementary substances that are so comparatively few in number, & still less different from one another in nature; it strongly suggests that, the further analysis can be pushed, the greater the simplicity, & homogeneity, that ought to be attained; thus, at length, we should have, as the result of a final decomposition, homogeneity & simplicity of the highest degree.[34]

Later this was elaborated:

> Now those things which are commonly called the Elements, Earth, Water, Air & Fire, are nothing else in my Theory but different solids and fluids, formed of the same homogeneous points differently arranged; & from the admixture of these with others, other still more compound bodies are produced.[35]

Boscovich's denial of contiguity is, in effect, an implicit recognition of the reality of space; for him it was not merely a well-founded phenomenon. In addition, and because they were homogeneous, the elementary points did not have unique 'souls' or spirits; very emphatically they were *not* perceiving monads:

> it is altogether false that there is no difference between my points & spirits. The most important difference between matter & spirit lies in the two facts, that matter is sensible & incapable of thought, whilst spirit does not affect the senses, but can think or will. . . . in these points of mine, I admit nothing else but the law of forces conjoined with the force of inertia; & hence I intend them to be incapable of thought or will.[36]

In differing from Newton, Boscovich pointed out that there must be a repulsive force if the distance between points became very small in order to prevent what he called compenetration.[37] But he agreed with Newton that matter must be in empty space and he argued that the void should not be thought of as separating the particles, rather the particles were to be thought of as floating in the void:

> The primary elements of matter are in my opinion perfectly indivisible & non-extended points; they are so scattered in an immense vacuum that every two of them are separated from one another by a definite interval; this interval can be indefinitely increased or diminished but can never vanish altogether without compenetration of the points themselves; for I do not admit as possible any immediate contact between them. On the contrary I consider that it is a certainly that, if the distance between two points of matter should become absolutely nothing, then the very same indivisible point of space, according to the usual idea of it, must be occupied by both together, & we have true compenetration in every way. Therefore indeed I do not admit the idea of vacuum interspersed amongst matter, but I consider that matter is interspersed in a Vacuum & floats in it.[38]

Boscovich was anxious to stress that even the most compact of bodies were still highly porous. This foreshadows our twentieth-century view of the atom (see chapter 6) as being largely empty space. Boscovich wanted utterly to reject the Cartesian view that matter was equivalent to extension (see chapter 3) and he also emphasised that so-called common sense could deceive and that it was necessary to interpret sensory evidence critically:

We obtain the idea of bodies through the senses; and the senses cannot in any way judge on a matter of accurate continuity; . . . Indeed we take it for granted that the continuity, which our senses meet with in a large number of bodies, does not really exist. In metals, marble, glass & crystals there appears to our senses to be continuity, of such a sort that we do not perceive in them any little empty spaces, or pores; but in this respect the senses have manifestly been deceived. This is clear, both from their different specific gravities . . . & also from the fact that several substances will insinuate themselves through their substance, For instance, oil will diffuse through the former, & light will pass quite freely through the latter . . .

. . . We may then suspect that accurate continuity without the presence of any little empty spaces – such as is certainly absent from bodies of considerable size, although our senses seem to remark its presence – is also nowhere existent in any of their smallest particles; but that it is merely an illusion of the senses, & a sort of figment of the brain through its not using, or through misusing, reflection. For it is a customary thing for men (& a thing that is frequently done) to consider as absolutely nothing something that is nothing as far as the senses are concerned; & this indeed is the source & principal origin of the greatest prejudices. Thus for many centuries it was credited by many, & still is believed by the unenlightened, that the Earth is at rest, & that the daily motions if the Sun & the fixed stars is proved by the evidence of the senses.[39]

As is evident from this passage, Boscovich also differed from Leibniz in holding that his non-extended points of energy were directly related to material bodies. Like space, they were part of ultimate reality and not merely well-founded phenomena. His aggregate of points possessed inertia*, they had mass and they could have momentum; he called the latter 'motion'.[40]

Boscovich expounded what Popper has called a *dynamic theory of extension* whereby extended matter is explained by something that is not matter, namely unextended points from which forces emanate.

The presence of (extended) matter in a certain region of space is a phenomenon consisting of the presence of repulsive forces in that region, forces capable of stopping penetration (or forces which are at least equal to the attractive forces plus the pressure at that place).[41]

Popper argues that Boscovich's theory, which he also connects with Kantian ideas (see below) was the 'direct forerunner of the Faraday–Maxwell theory of fields*.[42] Moreover, as Popper points out,[43] these primarily metaphysical speculations could be critically discussed and could be presented in a form that was subject to experimental test. It is important to appreciate this; Boscovich's theory was *more than* metaphysical speculation; it was an empirical theory which, in effect, proposed that the ultimate constituents of physical bodies were immaterial:

> *There is absolutely no argument that can be brought forward to prove that matter has continuous extension, & that it is not rather made up of perfectly indivisible points separated from one another by a definite interval; nor is there any reason apart from prejudice in favour of continuous extension in preference to composition from points that are perfectly indivisible, non-extended, & forming no extended continuum of any sort.*[44] [emphasis in the original]

Boscovich arrived at a theory as to the ultimate nature of matter which greatly influenced Sir Humphry Davy;[45] also it can readily be related to twentieth-century science. Heisenberg writes:

> The indivisible elementary particle of modern physics possesses the quality of taking up space in no higher measure than other properties, say colour and strength of material. In its essence, it is not a material particle in space and time but, in a way, only a symbol on whose introduction the laws of nature assume an especially simple form. . . . the experiences of present-day physics show us that atoms do not exist as simple material objects.[46]

KANT

Like Plato and Leibniz, Immanuel Kant (1724–1804) thought that ultimate reality the world of *noumena** (things in themselves) was not accessible to sense perception but, unlike them, he did not think that it was accessible to reason either. It might be immaterial or material we could know nothing about it. He accused Leibniz of constructing an *intellectual system of the world*,[47] believing himself to be 'competent to cognize the internal nature of things, by comparing all objects merely with the understanding and the abstract formal conceptions of thought'.[48]

Kant argued that in the phenomenal world there must be a permanent substratum underlying the changes we could observe:

> But the substratum of all reality, that is of all that pertains to the existence of things, is substance . . . the permanent, in relation to which alone can all relations of time in phenomena be determined, is substance in the world of phenomena, that is, the real in phenomena, that which, as the substratum of all change, remains ever the same.[49]

Substance could be known through the action of bringing about change, that is through exerting a force:

> Where action (consequently activity and force) exists, substance also must exist, . . . [Hence]
> . . . action alone, as an empirical criterion, is sufficient proof of the presence of substantiality.[50]

It is worth noting that this force is a phenomenal and therefore an observable (empirical) force, not some metaphysical postulate. Substances will act on each other and so Kant, like Boscovich, asserts that forces permeate space:

> every substance . . . must contain the causality of certain determinations in another substance, and at the same time the effects of the causality of the other in itself. That is to say, substances must stand (mediately or immediately) in dynamical community with each other, if co-existence is to be cognized in any possible experience . . .[51]

As Popper says, Kant's ideas influenced nineteenth-century theories of continuous fields of force and, like those of Boscovich, they can be seen as relating matter to energy.

SUMMARY

Plato (427–347 BC) thought that physical objects and materials were not proper objects of knowledge and that only the Forms (Universals) possessed true reality. This reality could be known only through thought. His views were influenced by Parmenides and by Pythagoras. Like the latter philosopher, he thought that variety in nature was the result of variety of geometric structure. For Plato material objects did exist, though they were but mere shadows of the Universals.

Leibniz (1646–1716) regarded material objects as well-founded phenomena, but not ultimately real. For him the underlying reality consisted of an infinite aggregate of immaterial souls, the monads. Monads were activated by entelechy, a vital force.

Boscovich (1711–87) thought that the ultimate basis of physical objects and materials consisted of immaterial points of force, with powers of repulsion and attraction. He had been influenced by Leibniz but also by Newton, and his theory of matter is closest to the twentieth-century view.

Kant (1724–1804) argued that in the phenomenal world substance (matter) had to be regarded as the permanent substratum in which change occurred. Substances had to stand in dynamic community with each other and therefore space must be permeated by forces.

3

From Atomos to Corpuscles

ULTIMATE PARTICLES AND THE INFLUENCE OF ARISTOTLE

In chapter 1 the ultimate particles suggested by Leucippus and Democritus were called atoms, and their theory was called an atomic theory. Even though the Ancient Greek 'atoms' had little in common with atoms as conceived in the twentieth century, the name has become so familiar that it would be pedantic not to use it when introducing those early theories. However, in this chapter, and from now on, I shall use the Ancient Greek word *atomos* to refer to their particle theories so that we shall not confuse them with those that came later. We need to consider how particle theories of matter (corpuscular theories) and also other theories as to the nature of matter developed.

For though from the seventeenth century until the late nineteenth century corpuscular theories triumphed there was another way of regarding the fundamental nature of matter, based on the Aristotelian concepts of matter and form. As we saw in chapter 1 (p. 14) Aristotle opposed atomos theories first, because he did not think it possible that there could be a void, and second, because he did not think that an atomos theory could account for changing qualities. He appreciated that a view of matter as being composed of discrete immutable particles was not compatible with his exposition of change in terms of matter as potentiality and of form as actuality (see chapter 1, p. 15). However, Aristotle was not opposed to the intellectual idea of *smallest parts*. He certainly entertained the notion of smallest parts of living organisms though he did not explicitly refer to the smallest parts of non-living matter. Nevertheless, whether they were of living matter or non-living matter, these *smallest parts*, or *minima*, were purely theoretical; Aristotle did not think that chemical or biological changes necessitated that the reacting material did in fact separate into smallest parts.

Those who espoused a corpuscular theory, on the other hand, held that the properties of a material were produced not as a result of a metaphysical (and mysterious) form being imposed on (an also mysterious) characterless matter, but rather by the arrangement of different kinds of atoms placed side by side. Those properties might change by the introduction of other atoms and/ or the rearrangement of atoms, just as a new word may be produced by different letters and/or the rearrangement of letters already there (see also chapter 1, p. 13). But at least up to the seventeenth century the Aristotelian idea of a change of form was more acceptable. It seemed at the time to provide a better explanation of the remarkable changes in quality that could occur when certain substances were heated together, or sometimes merely mixed together. Change of properties was explained by change of form, and change of form was simply an ultimate metaphysical fact; it was 'the way things were'.

MINIMA

Aristotle had used the notion of *minima* (see above) and this was developed in the early Middle Ages, but it was bedevilled by the problem of accounting for changes of property in terms of change of form. If elements were composed of minima and compounds were associations of different minima, how did the compound acquire its distinctive form? The Muslim Myrciana Avicenna of Bokhara (980–1037) suggested that the forms of the elements remained undiminished and integral, but that their properties changed through interaction with other elements. The Spanish Muslim Averroes (1125–98) who developed Aristotle's ideas thought that the form was debilitated; others suggested that the *minima* received some of the form of other *minima* whilst retaining some of their own. These explanations cannot be related to a corpuscular atomic theory since the physical *minima* were *not* held to be immutable; they were determined by their qualities, i.e. their form; they were subject to change because their form might change. Throughout the Middle Ages references to 'smallest parts' are references to *minima*, with changing properties, not to immutable atoms.

CORPUSCLES

By the seventeenth century explanations of change in terms of Aristotelian matter and form were complex and involved and they were being severely criticised. Other Ancient Greek ideas were revived. In chapter 2 we saw that though Plato believed that a shadowy matter existed, he argued that ultimate reality was non-material. This was also the view of Leibniz and Boscovich: for Leibniz, material bodies were well-founded phenomena which could indeed be subjects of physical inquiry; for Boscovich they were also subjects for chemical analysis. But many of their contemporaries, natural philosophers of the seventeenth and eighteenth centuries, regarded corporeal matter as a fundamental and ultimate reality, as fundamental as soul, mind or spirit.[1] They revived, modified and developed Greek atomosism in the form of corpuscular theories of matter.

We have seen that there were serious philosophical objections to corpuscular theories and twentieth-century views of matter are much closer to those of Boscovich, but it is a matter of historical fact that from the Renaissance until the close of the nineteenth century the various corpuscular theories contributed a great deal to the development of science and to the progress of scientific knowledge.

CORPOREAL PARTICLES AND THE VOID

It is possible to conceive of *non-corporeal* ultimate centres of force either as not in space at all since, in an ultimate sense, space is unreal (Leibniz), or as scattered through a void (Boscovich). But it would seem that any corpuscular theory must presuppose some space, albeit possibly minimal, between each corpuscle. For, with nothing between them, corpuscles would, in effect, be fused together into one vast, homogeneous Parmenidean plenum*. In addition, it might also be thought that if there were no space between bodies, then movement would be impossible. Hence, for this reason also, it would seem that there must be some space between ultimate particles. In fact, this second objection can be countered because in a plenum a swirling or circular movement can take place. As we shall see, Descartes proposed such motion; he also thought he had overcome objections to the possibility of corpuscles without space between them (see below).

PRIMARY AND SECONDARY QUALITIES

In chapter 1 (p. 13) it was shown how Democritus's atomos theory led to a distinction between primary qualities (those really in a material body) and secondary qualities (those resulting from bodies producing sensations in us). Despite the opposition of the Christian Church, based on its explanation of transubstantiation (see chapter 1, note 36), this view of sensible qualities as being dependent, directly or indirectly, on the properties of the ultimate corpuscles eventually came to replace the Aristotelian theory of matter and form as an explanation of physical properties. Not surprisingly a corpuscular explanation of sensible qualities developed along with corpuscular theories of the structure of matter in the seventeenth and eighteenth centuries, but there were different opinions as to what the primary qualities of matter *were*.

DESCARTES

René Descartes (1596–1650) propounded a corpuscular theory of matter even though he explicitly denied the existence of *ultimate* particles. He differed from Leibniz in that he asserted that corporeal matter was metaphysically real and not resolvable into forces or souls. But, like Leibniz (see chapter 2, quotation 19) he thought that any corporeal body was always potentially divisible. It follows that he discarded the notion of *minima* as well as any corpuscular theory:

> We see also the impossibility of atoms – pieces of matter that are by their nature indivisible. If they exist they must necessarily be extended, however small they are imagined to be; so we can still divide any one of them in thought (*cogitatione*) into two or more smaller ones, and thus we can recognise their divisibility.... Even if God made it not to be divisible by any creatures, he could not take away his power of dividing it; for it is quite impossible for God to diminish his own power.... So, speaking absolutely, it will still be divisible, being such by its very nature.[2]

Descartes' corpuscles were composed of the same basic matter.[3] They were *all* very small but they varied in size, smaller corpuscles having been formed by erosion from those which had been less small.

Descartes said that the only two essential properties of matter were extension and motion. At the Creation God had formed a definite and fixed quantity of matter and had endowed it with an absolute quantity of motion. It followed that the total quantity of matter and the total quantity of motion remained constant, though motion could be transferred so that individual corpuscles might sometimes be moving and sometimes be at rest. The sole primary qualities of the corpuscles and indeed of any corporeal body were extension and motion:

> The nature of matter, or of body considered in general, does not consist in its being a thing that has hardness or weight, or colour, or any other sensible property, but simply in its being a thing that has extension in length, breadth, and depth.
>
> ... weight, colour and all other sensible qualities of corporeal matter can be removed from body while it itself remains in its entirety; so it follows that its real nature depends upon none of them.[4]

Other qualities were secondary qualities:

> We therefore must on all counts conclude that the objective external realities that we designate by the words *light, colour, odour, flavour, sound,* or by names of tactile qualities such as *heat* and *cold,* and even the so-called *substantial forms* are not recognisably anything other than the powers that objects have to set our nerves in motion in various ways, according to their own varied disposition.[5]

Like so many philosophers, Descartes stressed that the senses could not be taken as a measure of what could be known,[6] and he contributed to the rejection of any explanation of qualities in terms of Aristotelian matter and form.

DESCARTES' DENIAL OF THE VOID

Descartes said that when a body expanded, and therefore became more rarefied, the space created between the original particles as they moved apart became filled with yet smaller particles so that there was never empty space.[7] For Descartes matter and extension were identical so that space was *equivalent to* matter; therefore, it was impossible for there to be a vacuum. Aristotle had said that nature abhorred a vacuum, but this implies that the non-existence of a void is merely contingent; Descartes held that the existence of empty space was *logically* impossible:

> The impossibility of a vacuum in the philosophical sense – a place in which there is absolutely no substance – is obvious from the fact that the extension of a space or intrinsic place is in no way different from the extension of a body. For the extension of a body in length, breadth and depth justifies us in concluding that it is a substance, since it is wholly contradictory that there should be extension that is the extension of nothing; and we must draw the same conclusion about the supposedly empty space – viz. that since there is extension there, there must necessarily be substance there as well.[8]

And again he says:

> there is a very close and absolutely necessary connexion between the concave shape of the vessel and the general concept of extension that must be contained in that concavity. It is no less contradictory than to think of a mountain without a valley, if we conceive that there can be this concavity without extension contained in it, or that there can be this extension without a substance whose extension it shall be . . .[9]

The argument is that if there were a void between two things, then there would be nothing between them and so they would be in contact with one another. As Williams points out, this is very like Parmenides's argument for a plenum.[10]

Williams contends that Descartes' view of matter as extension shows that his view of matter as an 'abstract geometrical conception' is inadequate and led to his 'too drastic assimilation of physics to those of pure mathematics'.[11] But Descartes thought he could use his metaphysical theory of matter as extension and motion to provide a physical explanation not only of the different properties of different materials but also (in due course) a complete account of the structure of the universe!

it is established that all bodies in the universe consist of one and the same matter; that this is divisible arbitrarily into parts, and is actually divided into many pieces with various motions; that their motion is in a way circular, and that the same quantity of motion is constantly preserved in the universe. We cannot determine by reason how big these pieces of matter are, how quickly they move, or what circles they describe . . . I shall suppose that all the matter constituting the visible world was originally divided by God into unsurpassably equal particles of medium size – that is of the average size of those that now form the heavens and stars; that they had collectively just the quantity of motion now found in the world; that . . . each turned round its own centre, so that they formed a fluid body, such as we now take the heavens to be; and that many revolved together around various other points . . . and thus constituted as many different vortices as there now are stars in the world. These few assumptions are, I think, enough to supply causes from which all effects observed in our universe would arise by the laws of nature . . .[12]

As will be shown, Descartes' contention that space (extension) was logically equivalent to matter and his rejection of the possibility of a vacuum were dismissed by many of his contemporaries and by those who came later. But though his theory of corpuscles swirling as vortices in a plenum was discarded,[13] he did introduce a new approach to inquiry which had lasting influence.

First, although he made many overt references to God in his discussion of the possibility of knowledge (God was therefore an essential postulate), in his consideration of the physical cosmos Descartes only 'used' God as the initial Creator of the cosmos – a seventeenth-century 'Big Bang'. He did not appeal to divine involvement or to any non-material force operating as part of the established laws of nature. Moreover his explanations were not dependent on biblical authority; in note 12 (above) he made no reference to the Genesis story. Admittedly, just prior to this passage he stressed that he was writing hypothetically,[14] but it is almost certain that this *caveat* was not sincere and was intended to protect him from Church censure and possible persecution. It is not surprising that his writings were put on the Index in 1660, a few years after his death.

Second, Descartes was the first to coin the phrase 'laws of nature', which he thought were ordained by God – the crucial point being that therefore they were absolute, i.e. they could not be broken. For him, as for most seventeenth-century and eighteenth-century natural philosophers, the aim of inquiry was to *discover* the laws of nature.[15] Although today we no longer regard laws of nature as divinely ordained and inviolable, our present aim of explaining the physical world in terms of such laws (which we would now think of either as observed regularities, or, more commonly, as deductions from our own theories) is largely developed from the Cartesian conception.

Third, Descartes was the first to liken the workings of natural objects to those of machines:

> The only difference I can see between machines and natural objects is that the workings of machines are mostly carried out by apparatus large enough to be readily perceptible by the senses . . . whereas natural processes almost always depend on parts so small that they utterly elude our senses. . . . it is just as 'natural' for a clock to tell the time, as it is for a tree grown from such-and-such a seed to produce certain fruit. So, just as men with experience of machinery when they know what a machine is for, and can see part of it, can readily form a conjecture about the way its unseen parts are fashioned; in the same way, starting from sensible effects and sensible parts of bodies, I have tried to investigate the sensible causes and particles underlying them.[16]

CARTESIAN DUALISM*

Although Descartes held that nearly all physical events could be explained on the basis of the mechanical interactions of matter according to the laws of nature, he was not a materialist. He thought that mind (soul or spirit) was also a fundamental reality: the essence of matter was extension, the essence of mind was thought. Because they were immaterial, minds were not subject to the laws of nature. Human beings were unique in being composed of mind and body[17] and their minds could direct their bodily actions.[18] This is a dualist theory and Descartes' particular exposition is called *Cartesian dualism*.[19] On this view human beings are not puppets subject to physical laws, and human actions are directed by the individual; therefore, each person is free to decide how to act and is morally responsible. He or she would be called to account on the Day of Judgement. Cartesian dualism dominated philosophy until the twentieth century and a form of it is still expounded by some distinguished scientists and philosophers today.[20]

ATOMICITY AND THE VOID

As indicated above Descartes' corpuscular theory – which in fact did much to revive atomism – is essentially self-contradictory since if matter is corpuscular, there *must be* space between the corpuscles. Francis Bacon (1561–1626) was more consistent; he gave limited support to a corpuscular theory, but said it was incomplete; there was no void but 'spirit' between the corpuscles. This reveals a latent animism, uninfluenced by Cartesian mechanism, though we should remember that in the sixteenth and seventeenth centuries there was no sharp distinction between the animate and inanimate; the term 'spirit' could also signify a tenuous fluid – what we would call a gas.

GASSENDI

Pierre Gassendi (1592–1655) was a little older than Descartes, but it was after Descartes had published his corpuscular theory that Gassendi developed his own theory. He introduced the notion of a molecule as a group of atoms. Gassendi opposed the Cartesian identification of matter with extension (space) and the consequent logical necessity of a plenum. It is worth noting that in 1645 Evangelista Torricelli (1608–47) discovered that the atmosphere would not support a column of mercury higher than about 30 inches and that a vacuum would be obtained above the column; this showed that nature did not necessarily abhor a vacuum, as Aristotle had supposed.

Gassendi argued that to picture space geometrically did not involve any appeal to the existence of matter. Rather, it entailed that matter be *distinguished from* the space in which it is found. For him the essential properties of matter were mobility (the possibility of movement), impenetrability and discontinuity. As

Koyré says, Gassendi was not a major figure and his corpuscular theory was that of the Ancient Greeks but his presentation

> enabled him on occasion to adopt ideas that were to have much success later on, as for example the corpuscular nature of light, of which, it must be confessed, he made no use (Newton was to do that), but even enabled him to surpass Galileo in formulating the principle of inertia, and Pascal in interpreting barometric phenomena.[21]

BOYLE

Robert Boyle (1627–91) also thought that matter was homogeneous and consisted of corpuscles in motion though, like Gassendi, he thought they must move in empty space. He also thought that each corpuscle was impenetrable and had a characteristic size and shape, so that impenetrability, size and shape were also essential properties of matter. But he agreed with Descartes[22] that the corpuscles must be mobile and that corpuscular motion must have been conferred by God:

> I agree . . . that there is one catholick or universal matter common to all bodies, by which I mean a substance extended, divisible and impenetrable.
>
> But because this matter being in its own nature but one, the diversity we see in bodies must necessarily arise from somewhat else than the matter they consist of. And since we see not how there could be any change in matter, if all its (actual or designable) parts were perpetually at rest among themselves, it will follow, that to discriminate the catholick matter into variety of natural bodies, it must have motion in some or all of its designable parts . . .[23]

Boyle did not always include impenetrability as an essential property,[24] but in other passages he also alludes to texture. All these were primary qualities and, like Descartes, he referred other qualities to interaction with our human sense organs:

> the body of man having several external parts, as the eye, the ear, etc. . . . . whereby it is capable to receive impressions from

the bodies about it, and upon that account is called an organ of sense; we must consider ... that these sensories may be wrought upon by the figure, shape, motion, and texture of bodies without them after several ways, some of those external bodies being fitted to affect the eye, others the ear, others the nostrils, etc. And to these operations of the objects on the sensories, the mind of man which upon the account of its union with the body,[25] perceives them, giveth distinct names, calling the one light or colour, the other sound, the other odour etc.... Whence men have been induced to frame long catalogues of such things ... we call sensible qualities; and because we have been conversant with them before we had the use of reason ... we have been from infancy apt to imagine that these sensible qualities are real beings in the objects they denominate.... whereas indeed ... there is in the body, to which these sensible qualities are attributed, nothing of real and physical but the size, shape, and motion or rest, of its component particles.[26]

Boyle also agreed with Descartes that, *in principle*, there was no theoretical limit to the divisibility of corpuscles but, unlike Descartes, he thought that *in fact* corpuscles were *minima naturalia* and were indivisible.[27] Interestingly, he also developed the idea that groups of corpuscles could adhere into relatively stable clusters,[28] which could be the basis of different materials. Moreover, alterations in the clusters would bring about changes in appearance; these could be reversible, but there might be a major change in properties:

we see that even grosser and more compounded corpuscles may have such permanent texture: for quicksilver, for instance, may be turned into a red powder[29] or a fusible and malleable body, or a fugitive smoke,[30] and disguised I know not how many other ways, and yet remain true and recoverable mercury. And these are, as it were the seeds or immediate principles of many sorts of natural bodies ...

That as well each of the *minima naturalia*, as each of the primary clusters ... having its own determinate bulk and shape, when these come to adhere to one another, it must always happen that the size and often the figure of the corpuscle composed by their juxtaposition and cohesion will be changed; ...

And whether anything of matter be added to a corpuscle or taken from it . . . the size of it must necessarily be altered, and for the most part the figure will be so too, whereby it will acquire a congruity to the pores of some bodies . . . and become incongruous to those of others; and consequently be qualified . . . to operate on divers occasions, much other wise that it was fitted to do before.[31]

Boyle's theory is a corpuscular theory rather than an atomic theory; he envisaged the complex corpuscles which he called *aggregates* or *primary concretions* behaving as chemical elements. Theoretically, these concretions could be broken up, but in practice it was difficult to do this. The corpuscles were like Gassendi's molecules. For both Boyle and Gassendi, the notion of an *ultimate particle*, or atom, was redundant.

It might be thought that Boyle's corpuscles and Gassendi's molecules were analogous to our present concept of molecules, i.e. a unit composed of atoms, but the analogy is false. First, it appears that Boyle's corpuscles and Gassendi's molecules were not so readily decomposed as today we take our molecules to be. Second, it seems that Boyle, at least, thought that it was the arrangement of the *corpuscles* (not the arrangement of the particles which made up the corpuscles) that gave the substance its distinctive physical properties. He did recognise that chemical properties could not be accounted for by a purely mechanical arrangement, but he was hopeful that in time this would be possible. Of course, like Descartes, he absolutely rejected the Aristotelian doctrine of forms.

Appealing to his corpuscular theory Boyle was able to account for physical laws, such as his own well-known relation of gas pressure and volume and also for various chemical changes, such as the increase in weight of metals on calcination*. Many of his explanations are rejected today: for example, he thought the weight gained in calcination was due to the absorption of particles of fire, but Boyle's corpuscular explanations could be tested by further experiment (as, in the eighteenth century, was Lavoisier's explanation of the weight gain after calcination). Even though Boyle's explanations were wrong, they represented progress in scientific method in that they *were* subject to critical appraisal and test, i.e. they were refutable and therefore genuine empirical theories.

There is no doubt that Boyle regarded mechanical explanations, based on the movements of corpuscles, as eminently satisfying:

> And here we have a fair occasion to take notice of the fruitfulness and extent of our mechanical hypothesis: for since, according to our doctrine, the world we live in is not a moveless or indigested mass of matter, but [a] *self-moving engine*, wherein the greatest part of the common matter of all bodies is always . . . in motion, and wherein bodies are so close set by one another, that . . . they have either no vacuities betwixt them, or only here and there interposed and very small ones . . . it will very naturally follow, that from the various occursions of those innumerable swarms of little bodies that are moved to and fro in the world, there will be many fitted to stick to one another and to compose concretions; and many disjoined from one another and agitated apart it will not be hard to conceive that there may be an incomprehensible variety of associations and textures of the miniute parts of bodies, and consequently a vast multitude of portions of matter endowed with store enough of differing qualities.[32]

Very reasonably, Boyle argued a case for alchemy and neither he nor Newton ruled out the possibility of transmutation:

> So that though I would not say that any thing can immediately be made of every thing, as a gold ring of a wedge of gold or oil or fire of water; yet since bodies having but one common matter can be differenced but by accidents, which seem all of them to be the effects and consequents of local motion . . . I see not why it should be absurd to think that (at least among inanimate bodies) by the intervention of some very small addition or subtraction of matter . . . and of an orderly series of alterations, disposing by degrees the matter to be transmuted, almost any thing, may at length be made anything.[33]

LOCKE

John Locke (1632–1704) also adopted a corpuscular theory of matter; he thought it the best available, though he was well aware that it was hardly adequate to account for the observed qualities of bodies:

I have here instanced in the corpuscularian hypothesis, as that
which is thought to go furthest in an intelligible explication of
the qualities of bodies; and I fear the weakness of human understanding is scarce able to substitute another which will afford
us a fuller and clearer discovery of the necessary connexion
and *co-existence* of the powers which are to be observed united
in several sorts of them. This at least is certain: that whichever
hypothesis be clearest and truest (for that it is not my business
to determine),[34] our knowledge concerning corporeal substances
will be very little advanced by any of them, till we are made
to see what qualities and powers of bodies have a *necessary
connexion or repugnancy* one with another; which in the present
state of philosophy[35] I think we know but to a small degree.[36]

Explanations of 'connexion and repugnancy' (we would now say
'attraction and repulsion') were to be offered at the end of the
eighteenth century with investigations into electricity (see chapter 6), but in Locke's time attraction and repulsion had to be
accepted as brute facts. Here Locke used the corpuscular hypothesis
to try to account for a change of state from water to ice:

The little bodies that compose the fluid we call *water* are so
extremely small that I have never heard of anyone who by a
microscope (and yet I have heard of some that have magnified
to 10,000; nay to much above 100,000 times) have pretended
to perceive their distinct bulk, figure, or motion; and the particles of *water* are also so perfectly loose one from another that
the least force sensibly separates them. Nay, if we consider
their perpetual motion, we must allow them to have no cohesion
one with another; and yet let but a sharp cold come and they
unite, they consolidate, these little atoms cohere and are not, without
great force, separable. He that could find the bonds that tie these
heaps of loose little bodies together so firmly, he that could
make known the cement that makes them stick so fast one to
another, would discover a great and yet unknown secret.[37]

Locke also used the corpuscular theory to account for the properties of bodies. Like Boyle he allowed more than extension and
motion to be primary qualities; he distinguished secondary qualities
as powers in much the same way as Descartes:

First, the *ideas* of the primary qualities of things, which are discovered by our sense, and are figure [shape], number, situation and motion of the parts of bodies, which are really in them, whether we take notice of them or no. *Secondly*, the sensible secondary qualities, which depending on these, are nothing but the powers those substances have to produce several *ideas* in us by our senses; which *ideas* are not in the things themselves otherwise than as anything is in its cause.[38]

As Jackson has pointed out,[39] in the strict sense Locke took only primary qualities to be qualities, the secondary qualities *of bodies* being, in effect, just powers to produce sensations in us. They need to be distinguished from *sensible* secondary qualities which are the actual ideas of colour, odour, temperature, etc. that we experience. Locke held that these ideas (sensible secondary qualities) were all caused by the primary qualities of the corpuscles – the insensible parts.[40] We have seen that Locke was aware that there was still much to explain, but he thought the best possibility of further understanding would come by developing a corpuscular hypothesis.

Notes 34 and 35 of this chapter show that in the seventeenth century there was no distinction between what we now call science and philosophy; Boyle, Locke and Newton would all have called themselves natural philosophers. But we can also see that by his caveat in note 36 ('it is not my business to determine'), Locke appreciated that his philosophical speculations were less closely connected with empirical inquiry than were those of many of his contemporary natural philosophers. We have seen how Boyle used the corpuscular hypothesis to account for certain observed physical and chemical changes.

NEWTON

Isaac Newton (1642–1727) did not directly use the corpuscular theory in his mechanics, though it is there by implication (see below). He was concerned with the physical behaviour of matter at the macro-level: with the concepts of mass, extension, density, force and weight and with ways of relating them to mathematically formulated laws of nature.

He made use of a corpuscular theory in his account of light[41] (which eclipsed Huygens's wave theory*) and also in his alchemical investigations (see reference to Boyle, p. 47). Like Descartes, he believed that the corpuscles had been created by God, but he thought that in addition to extension and mobility the corpuscles' primary qualities were also mass, solidity (hardness and impenetrability) and figure (shape):

> It seems probable to me, that God, in the Beginning form'd Matter in solid, massy, hard, impenetrable, moveable Particles, of such Sizes and Figures, and with such Properties, and in such Proportion to Space, as most conduced to the End for which he form'd them.[42]

For Newton, the basic corpuscles were of constant density[43] and were the ultimate constituents of matter; they had objective reality and they (and their primary qualities) existed independently of our sense experience. The mass of a macro-object was the sum of the masses of its constituent corpuscles, and hence all material entities had mass, and mass was a primary quality of macro-objects as well as of the minute corpuscles. Newton's use of the notion of inertial mass* and his concepts of force (in particular of weight) presuppose the fundamental and objectively real quality of mass that the corpuscular theory of matter postulates. Moreover, his view was that matter was not passive: there were inherent forces between corpuscles (see below) and the notion of inertial mass implied a force, *vis inertiae* in matter itself.

Somewhat like Boyle, Newton thought that the basic corpuscles (essentially atoms) cohered in clusters and that such clusters would group to form clusters of clusters 'and so on for divers successions, until the progress end[s] in the biggest particles on which the operations in chymistry, and the colours of natural bodies depend, and which by cohering compose bodies of sensible magnitude.'[44] All macro-matter had a hierarchical structure and was essentially compound – hence the possibility of transmutation. Since the basic corpuscles had the same density, differences in the density of materials were due to differences in the proportion of empty space both between clusters and within clusters: even a dense material such as gold must be largely empty space. It followed that basic corpuscles were almost certainly very very dense and that 'all the solid matter in the solar system might be contained within a nut-shell'.[45]

Newton's concept of very dense ultimate particles might appear to be directly related to twentieth-century views of enormously dense matter composed of atomic nuclei, but the connection is misleading since in this century we do not take ultimate particles as having inertial mass in Newton's sense. However, Newton's view of high density matter and of forces within matter did influence eighteenth-century thought and helped to lead Boscovich to the view that passive matter was redundant. As we saw in the previous chapter (p. 32) with Boscovich there is indeed a link to twentieth-century science.

Yet, despite his great prestige, development of a corpuscular theory of matter did not come through the Newtonian concept. In the eighteenth century there were religious doubts since Newton's rejection of the notion of passive matter paved the way for a rejection of dualism with its distinction of matter and spirit. Joseph Priestley (1733–1804), a supporter of Boscovich's development of Newton's theory, argued for a monist view of matter and spirit, and this was one reason why Priestley was persecuted: such a monism was regarded as rank materialism. Boscovich, a Jesuit priest, of course rejected materialism and so did the Quaker John Dalton.[46]

Hence it was the less sophisticated corpuscular theory of matter (sub-microscopic hard particles moving through empty space) that was supported by most natural philosophers, and by the mid-eighteenth century it was accepted even in France, where the Cartesian theory had been dominant. It could be used to explain many physical properties of material objects: their sensible qualities (primary and secondary) and much of their mechanical behaviour (inertia, elasticity, recoil or change of direction on impact). But explanation of the latter was not logically dependent on a corpuscular theory of matter. In addition – and this was a defect as compared with Aristotelian theories of matter and form – a simple corpuscular theory did not adequately account for the dramatic changes of properties which could occur in chemical reactions. However, since all the various corpuscular theories postulated one kind or only a few kinds of ultimate particles, which in some way could produce different materials, they were all compatible with alchemical theories of transmutation. As has been noted, both Boyle and Newton believed in the possibility of transmutation. By the mid-eighteenth century, albeit Aristotelian ideas were discarded, the corpuscular theory, though providing good explanations of certain physical changes, was of little value to the chemist. It could not be a basis for useful predictions as to the outcome of chemical reactions, and it could not provide any explanation of chemical change beyond the bare statement that corpuscles must be rearranged.

There were also purely philosophical objections which led to questioning of the concept of objectively real primary qualities and of secondary qualities as powers. These will be considered in chapter 5, but for the moment we shall pursue the progress of the corpuscular theory as a basis for scientific inquiry into the nature of matter and the processes of chemical change.

SUMMARY

In the seventeenth century the Aristotelian account of change in terms of matter and form was becoming unacceptable and corpuscular theories of matter were developed.

Descartes (1596–1650) postulated varying sizes of corpuscles moving in a plenum; he argued that a vacuum was a logical impossibility. He distinguished two primary qualities of matter – extension and motion – from secondary qualities. Secondary qualities arose as a result of the corpuscles acting on our senses. Descartes also coined the phrase 'laws of nature' and likened the cosmos to a vast machine. Human beings were the only physical entities exempt from causal laws, because they were animated by an immaterial mind (soul or spirit). This is Cartesian dualism.

Later philosophers disagreed with Descartes and held that corpuscles moved in empty space (the void), Gassendi (1592–1655) and Boyle (1627–91) also discussed the notion of complex corpuscles, analogous in some (but not all) ways to the nineteenth-century concept of molecules.

Locke (1632–1704) accepted a corpuscular theory and discussed attraction and repulsion between corpuscles as well as primary and secondary qualities.

Newton (1642–1727) developed a theory of basic homogeneous corpuscles endowed with an inherent force – *vis inertiae*. He argued that macro-matter must be made up of a hierarchy of clusters of basic corpuscles and that it would consist largely of empty space.

But by the eighteenth century a more simple corpuscular theory of matter was generally accepted. Such a theory could explain physical properties but, beyond the bare statement that corpuscles must be capable of rearrangement, the corpuscular theory could not account for chemical changes.

4

From Corpuscles to Atoms and Molecules

In chapter 3 we saw that though the term 'natural philosophy' was applied both to what, today, we call 'philosophy' *and* to what we call 'science', a natural philosopher such as Locke was aware that there were different branches to his subject. By the seventeenth century there was, in practice, a marked distinction between those natural philosophers whom we would regard as scientists, such as Priestley, Cavendish and Lavoisier, and other philosophers, such as Berkeley and Hume. We shall see that their approaches to the problem of matter were very different.

CORPUSCLES AND CHEMISTRY

In this chapter we shall consider the scientific approach and how changes occurring when materials are put together, and especially when they are heated (chemical changes), were studied and explained by scientists. We shall also consider the extent to which those explanations came to be related to corpuscular theories. Corpuscular explanations culminated in Dalton's atomic theory, a theory that was firmly based on appeal to observation and experiment. By contrast with the Greek atomos theories, it is tempting to assert that Dalton's theory was altogether free from metaphysical speculation, but this would be misleading, indeed incorrect. This is not to disparage Dalton for there cannot be a complete separation of metaphysical ideas from science; all scientific theories, including chemical theories of the nature of elements and compounds, incorporate metaphysical along with philosophical assumptions. Metaphysical assumptions must be there, though the distinctive feature of any theory rated as a scientific theory is that it is subject to test from observation and possibly planned experiment.

As we have seen, even early theories of matter which many tend to regard as largely imaginative fantasies have had some support from observation. Classical atomos theories, seventeenth-century corpuscular theories and eighteenth-century theories of points of force may seem, superficially, to be primarily metaphysical/philosophical speculation, but they have a claim to be taken as genuine scientific (empirical) theories since they were used to explain what was observed.'[1] Hence though it has to be admitted that they were not as firmly empirical as Dalton's atomic theory and are often treated as philosophical[2] rather than scientific theories, they do have some scientific respectability; they were not 'deduced' from metaphysical postulates asserting what the world *must* be like. In any case, and this must be stressed, scientific theories can never have the certainty of mathematical truths. We hope that observations will confirm and not refute an explanatory theory but we can never aspire to a logical proof. This holds for all scientific theories, including those currently accepted and regarded as firmly established. Thus though observations (and experiments) may provide good grounds for accepting or for rejecting an explanatory theory, they can never logically demonstrate its truth or its falsity.[3]

By the late eighteenth century, chemists were operating within a framework of theories analogous to the Newtonian framework for physics. Terms such as 'element' and 'compound'* acquired an accepted significance and certain substances became established as elements. There were also theories as to the nature of heat and combustion.[4] Some of these theories were discarded, but a consequence of the theoretical framework was that scientists could plan experimental tests and were no longer amassing observational evidence in a haphazard, almost cookery book manner. Again, by the start of the nineteenth century the importance of weighing reactants and products of chemical changes was firmly established and methods of weighing were continually improved. Descriptions of reactions were not merely qualitative accounts of changes in properties, they were quantitative accounts of changes in weight. In many cases it also became possible to obtain pure materials or, if they were not pure, it was possible to be aware of this and to have some measure of the extent of impurity. It was because it could explain the new quantitative laws of chemical combination that Dalton's atomic theory had much greater empirical support than had earlier corpuscular theories. Dalton's theory typifies what we may call the scientific (as opposed to the philosophical) approach to the problem of the mystery of matter.

But before we consider the Daltonian development of seven-teenth-century corpuscular theories, we need to assess the development of general views of chemical changes and theories as to how they occurred.

COMBUSTION AND CHEMICAL CHANGE

It was, of course, appreciated that many chemical changes could be brought about by heat and there was much interest in the process of combustion. Leonardo da Vinci (1452–1519) had stated that burning that was accompanied by a flame would not take place without air and, from the sixteenth century onwards, it was generally accepted that air played an important part in combustion and that there was an analogy between combustion and respiration. The problem of what was *happening* during combustion started to be investigated methodically in the seven-teenth century. Jean Rey (fl. 1630) proposed that the increase in weight of metals on calcination was due to air *mixing with* the burning substance, but his proposal was ignored or denied. It may be that it was thought unimportant, or was rejected, because it was known that greater heat (which encouraged more vigor-ous combustion) led to a greater increase in the weight of the calx, whereas if Rey were right, greater heat would be expected to drive out some of the air, thereby leading to less of an increase in weight. In addition Rey's theory took no account of change of properties (qualities) during combustion and for that reason it tended to be disregarded by chemists who were still dominated by the alchemists' conception of qualities, based on Aristotelian doctrines.

However, we must also bear in mind that, at the time, arguments based on considerations of weight change or weight conserva-tion were not regarded as convincing. Thus in the sixteenth century it was held that only coarse matter (large lumps) had mass. Descartes, for example, did not think that weight was an immu-table property of matter. Neither Descartes nor Leibniz distin-guished mass from weight. As late as 1808 (and despite the prestige and influence of both Lavoisier and Dalton; see below) there were still some lingering doubts about the invariability of mass in chemical change. Nevertheless, through Lavoisier's championship, conservation of mass (or weight) in chemical change did eventually become adopted as a basic *a priori* principle, though it was not until the early twentieth century that the principle was supported by really accurate weighing. It was then shown that any weight change would be less than one part in 10,000,000.[5]

Robert Hooke (1635–1703) noted that there was something in potassium nitrate* which allowed burning to occur, and he suggested that it was the same substance which permitted burning in air. He thought of combustion as a sort of dissolving in air. John Mayow (1645–79) had also noted that there was something in gunpowder which encouraged burning and offered an explanation of combustion based on the supposition that there were nitro-aerial particles mixed with air corpuscles. He suggested that such particles could corrode metals and would react with flammable material; he also argued that in respiration they were absorbed and would alter the colour of blood. It might seem that Mayow had preceded Lavoisier and was an early advocate of an oxygen theory of combustion and respiration, but this is doubtful since he appealed to nitro-aerial particles to account for a great many other unrelated changes. However, his use of a particle theory showed how such theories could account for many chemical changes in terms of accretion or loss of constituent corpuscles from a complex corpuscle, and/or from their rearrangement.

Boyle showed that when substances were heated they gained weight, but he did not accept Rey's explanation that the increase in weight came from the air; he thought it came from the heat. Boyle heated tin in a vessel which had previously been partly evacuated before sealing and weighing. (The evacuation was made to prevent the tin from bursting as the heat would have made air inside expand.) After heating, Boyle weighed the tin again; but the second weighing was done *after* the seal had been broken and of course there was an increase in weight due to the inrush of air equalising the pressures. Because Boyle did not notice this inrush of air he thought that the weight increase came from the heat. On the basis of this work and other observations, Boyle believed that metals were elements and that the calx* (what we now call the oxide) was a compound of the metal and fire.

Also in the seventeenth century, there arose a very different theory of combustion: Joachim Becher (1635–82) had suggested that all combustible materials contained an oily earth, *terra pinguis*, which was responsible for their being flammable. Later, Georg Ernst Stahl (1660–1734) developed Becher's hypothesis, calling *terra pinguis* 'phlogiston' and treating it as a principle of fire. Stahl said that when materials burned, phlogiston escaped in the flame. His theory of combustion dominated mid-eighteenth-century chemistry but, as with contemporary approaches to the problem of affinity, it did not depend on a corpuscular hypothesis.

The phlogiston theory was reluctantly discarded and by the end of the century it slowly became accepted that air was a mixture of gases, one of which was oxygen, and that combustion (and respiration) depended on chemical reaction with oxygen[6] in the air. It was beginning to be realised that chemical compounds were distinct from mixtures, and that chemical changes needed to be distinguished from mere mixing.

Also, Bacon's suggestion of heat as motion was revived[7] and began to replace the caloric theory of heat. However, the caloric theory could be used to provide a satisfying explanation of the compressibility of elastic fluids (gases) and it lingered on till the mid-nineteenth century. Indeed, Dalton and some of his contemporaries based explanations of chemical affinity on the assumption that there were particles of heat. At the start of his *A New System of Chemical Philosophy* Dalton wrote:

> The most probable opinion concerning the nature of caloric is, that of its being an elastic fluid of great subtilty, the particles of which repel one another, but are attracted by all other bodies.[8]

DALTON'S SEVENTEENTH- AND EIGHTEENTH-CENTURY PRECURSORS

In the late sixteenth century Daniel Sennert (1572–1637) accepted a corpuscular theory of matter and thought there were specific differences between corpuscles of different materials. But he made a clear distinction between simple corpuscles and more complex bodies, those of what we should now call chemical compounds;[9] Sennert

> spoke of atoms whose minuteness could be adduced from the penetrating powers of vapours, from the small size of perfectly formed organisms like insects, from the differences in nature

between a liquid and its vapour, and from the variety of changes through which substances like metals can pass and yet be recoverable in their original form.[10]

We have seen that Boyle used the corpuscular theory to explain chemical and physical changes which, like Sennert, he had himself observed.

> I hoped I might at least do no unreasonable piece of service to the corpuscular philosophers by illustrating some of their notions with sensible experiments, and manifesting that the things by me treated of may be at least plausibly explicated without having recourse to inexplicable forms, real qualities, the four peripatetick[11] elements, or so much as the three chemical principles.[12]

Boyle has been credited with introducing a pragmatic definition of an element, as something which could not be analysed into more simple constituents. It is true that his criticisms of earlier element theories show he did not accept accounts derived from Aristotle or other 'authorities' (see quotation above), but the pragmatic definition of an element was to come a century later from Antoine-Laurent Lavoisier (1743–94) in his *Elements of Chemistry* (1789) – an element being 'the actual term whereat chemical analysis has arrived'. Hall argues that Boyle's corpuscular theory would have actually prevented him arriving at the pragmatic definition. For Boyle:

> No body made up of corpuscles could be simple homogeneous or elementary in the strict sense, because the corpuscles themselves were compounds of various particles. The pragmatic test of resistance to analysis proved only that some concretions of these particles into corpuscles were more coherent than others, whether brought about by nature (gold) or art (glass). ... the truth that chemical changes occur by modification of corpuscular structure and be re-shuffling of the particles within corpuscles, was far more significant to him than a dubious search for for primary elements, whose existence he was quite content to regard as hypothetical.[13]

COMPOUNDS AND AFFINITY

Nicolas Harsoeker (1656–1725) offered a corpuscular theory whereby attraction and repulsion were related to the shape of corpuscles:

> According to him round particles slide over each other in liquids, sharp particles cohere in solids. Salts and vitriols have complex particles, heavy balls bearing sharp blades or needles. When a metal dissolves in acid, the acid particles penetrate between the metal particles and then water particles penetrate into the gap opened up.[14]

But a major problem for those who supported a corpuscular theory of chemical reaction and chemical change was to offer an explanation as to how the corpuscles came to adhere, in many cases so very firmly, and yet also how they came to separate; this was the problem of chemical affinity. It was generally believed that heat was an element[15] and some thought (as did Boyle, see above) that there were particles of fire and that a gravity-like attraction[16] between different material corpuscles might be reduced by fire particles which would cause repulsion.[17] It was surmised that the compressibility of gases might be due to a compressible globular envelope of caloric* (heat) surrounding each corpuscle. Boyle's gas law was explained by supposing that through its agency the chemical units repelled each other.[18] Palmer quotes directly from Dalton:

> When any body exists in the elastic state, its ultimate particles are separated from each other to a much greater distance than in any other state; each particle occupies the centre of a comparatively large sphere, and supports its dignity by keeping all the rest, which by their gravity, or otherwise, are disposed to encroach up to it, at a respectful distance.[19]

At this time the concept of heat as a mode of motion, which was to be developed by Maxwell and others into a kinetic theory of gases, was not generally recognised, so that expansion and contraction were understood in a relatively static sense.

In 1718 Étienne-François Geoffroy (1672–1731) had published tables of chemical affinity which remained an authoritative reference throughout the eighteenth century. They had been supported by experimental evidence but later they were shown to be inadequate. Most chemists accepted that matter might be composed of tiny ultimately indivisible particles but in the mid-eighteenth century, despite the hypothesis of gravity-like attraction and fire-particle repulsion, corpuscular theories were not judged to be helpful in accounting for particular cases of affinity. Many chemists, like Lavoisier, thought that theories as to the nature of matter should be disregarded; it was only necessary to know what substances were elements. Lavoisier acknowledged that an atom could be thought of as the smallest unit appearing in a chemical reaction, but this did not entail that the unit was indivisible. Yet Lavoisier's assumption of the principle of the conservation of matter was soon to be supported by Dalton's atomic theory. It is true that the principle of the conservation of matter does not presuppose an atomic theory but, given an atomic theory, the principle of conservation of matter follows.

QUANTITATIVE EVIDENCE FOR DALTON'S ATOMIC THEORY

During the eighteenth century the value of quantitative measurement came to be appreciated and some important quantitative chemical laws were established. In 1791 Jeremiah Richter (1762–1807) formulated the law of equivalent proportions,[20] and in 1797 Joseph-Louis Proust (1754–1826) stated the law of constant composition. Both these laws were, of course, based on experimental investigations; they were not purely philosophical or metaphysical speculations.

John Dalton (1766–1844) arrived at his theory from a study of gas pressures, particularly the pressures of the constituents of the atmosphere, which he thought, as did many of his contemporaries, was a mixture of gases. But he had to explain why it was that, if air were a mixture, the corpuscles did not separate into layers according to their specific gravities*. He suggested that though atoms did not repel other atoms of different kinds, they did repel atoms of their own kind. It was for this reason that Dalton would not accept that atoms of the same kind could combine together in molecules.

However, Dalton appealed to Richter's and Proust's quantitative laws to support his theory, as we shall see in his exposition as given below. This first appeared in a paper given in 1803 and published in 1805, but it was more explicitly formulated in his *New System of Chemical Philosophy* published in 1808. He postulated:

1. That the atoms of a simple substance, i.e. an element, were all identical, but different from the atoms of any other element. They differed in size and in number per unit volume and, most importantly, they differed in weight. This postulate is supported by the law of constant composition.
2. That when two elements combined to form a compound atom each atom of the first element united with one atom, or a small whole number of atoms, of the second element. This is supported by the law of equivalent proportions.
3. If more than two compounds could be formed from two elements there was a simple ratio between the combining weights. This was Dalton's law of multiple proportions and, as with the other laws, it was supported by experimental evidence. Thus the atomic theory accounted for this law just as it accounted for the law of equivalent proportions and the law of constant composition.

Of course, Dalton could not weigh individual atoms but he pointed out that atoms could be characterised by their *relative weights*.[21] He made the assumption that hydrogen (the least dense gas) was composed of the lightest atoms and he also made the assumption that the combining ratios of different atoms would be the most simple.[22] For example, he assumed that the complex atom of water was made of one hydrogen and one oxygen atom; experimental analysis then showed that each oxygen atom was eight times heavier than each hydrogen atom.[23]

We know now that Dalton was right about hydrogen, though some of his other assumptions – such as the assumption as to the composition of water – were incorrect. However, the importance of his theory is that it transformed the earlier speculative, albeit empirical, corpuscular hypotheses (empirical because they could indeed account for qualitative changes) into a quantitative theory which could be supported by (or refuted by) chemical analysis. It is interesting to see that Dalton made no Cartesian reference to God and the Creation, though he fully appreciated the importance of the postulate[24] that the total amount of matter could not be changed. Quantitative evidence would be worthless unless that postulate were accepted. In the *New System of Chemical Philosophy* he wrote:

> Chemical analysis and synthesis go no farther than to the separation of particles one from another, and their reunion. No new creation or destruction of matter is within the reach of chemical agency. We might as well attempt to introduce a new planet into the solar system or annihilate one already in existence; as to create or destroy a particle of hydrogen. All the changes we can produce, consist in separating particles that are in a state of cohesion or combination and joining those that were previously at a distance.
>
> In all chemical investigations, it has justly been considered an important object to ascertain the relative *weights* of the simples which constitute a compound. But unfortunately the inquiry has terminated here; whereas from the relative weights in the mass, the relative weights of the ultimate particles or atoms of the bodies might have been inferred, from which their number and weight in various other compounds would appear. . . . Now it is one great object of this work, to show the importance and advantage of ascertaining *the relative weights of the ultimate particles, both of simple and compound bodies, the number of simple elementary particles which constitute one compound particle, and the number of less compound particles which enter into the formation of one more compound particle.*[25] (emphasis in the original)

Greenaway says that 'the paradox of Dalton's theory was that it concentrated attention on weight determination when all the time it was capable of illuminating an aspect of the constitution of matter to which the weight of atoms was really irrelevant.'[26] However, it must be borne in mind that the weight laws cited above gave very important, indeed vital, empirical support for Dalton's theory and it is not surprising that his exposition gave prominence to the quantitative evidence already available.

Though others might ignore the structural implications, Dalton himself and his contemporary, William Hyde Wollaston (1766–1828), were aware that the atomic theory could help in discovering the structure of matter. In a paper read to the Royal Society in 1808 Wollaston wrote:

I am . . . inclined to think, that when our views are sufficiently
extended, to enable us to reason with precision concerning the
proportions of elementary atoms, we shall find the arithmetical relation alone will not be sufficient to explain their mutual
action, and that we shall be obliged to acquire a geometrical
conception of their relative arrangement in all three dimensions of solid extension.[27]

Further, Wollaston saw that structure might also give some clue
as to why elements combined and separated:

It is perhaps too much to hope that the geometrical arrangement of primary particles will ever be perfectly known; . . . until
it is ascertained how small a proportion the primary particles
themselves bear to the interval between them, it may be
supposed that surrounding combinations, although themselves
analogous, might disturb that arrangement, and in that case,
the effect of such interference must also be taken into account,
before any theory of chemical combination can be rendered
complete.[28]

In his Bakerian lecture of 1812, Wollaston assumed that there
were 'elementary particles' and showed how crystalline structure could be accounted for by assuming that they were perfect
spheres. He pointed out that Hooke took the same view in his
seventeenth-century *Micrographia*, where he referred to 'globular
particles'.[29]

Nevertheless, from 1820 to 1860 the atomic theory did not play
a large part in chemical explanations. The situation was analogous
to the attitude to corpuscular theories a century before, but
difficulties were enhanced because certain nineteenth-century
hypotheses arising from and supported by experimental discoveries
could not be explained on the basis of the atomic theory without
contradictions and inconsistencies. Also there was conflicting
experimental evidence in establishing certain atomic weights. In
short, there were so many lacunae, so much inconsistency and,
in addition, some doubts about experimentally determined weights,
that many chemists were sceptical as to the actual existence of
atoms and preferred to deal with the combining weights (equivalent weights*) of elements and to set atomic speculation to one
side. Thus as early as 1814, Wollaston adopted a new approach

having decided that it was more practicable to work with equivalent weights rather than with atomic weights:

> According to Mr. Dalton's theory, by which these facts are best
> explained, chemical union in the state of neutralization* takes
> place between single atoms of the substances combined; and in
> cases where there is a redundance of either ingredient, then two
> or more atoms of this kind are united to only one of the other.
>
> According to this view, when we estimate the relative weights
> of equivalents, Mr. Dalton conceives that we are estimating the
> aggregate weights of a given number of atoms, and consequently
> the proportion which the ultimate single atoms bear to each
> other. But since it is impossible in several instances, where only
> two combinations of the same ingredient are known, to discover
> which of the compounds is to be regarded as consisting of a
> pair of single atoms, and since the decision of these questions
> is purely theoretical, and by no means necessary to the forma-
> tion of a table adapted to most practical purposes, I have not
> been desirous of warping my numbers according to an atomic
> theory, but have endeavoured to make practice my sole guide,
> and have considered the doctrine of simple multiples, on which
> that of atoms is founded, merely as a valuable assistant in
> determining, by simple division, the amount of those quanti-
> ties that are liable to such definite variations from the original
> law of Richter.[30]

Again, in 1826 when Sir Humphrey Davy (1778–1829) as Presi-
dent of the Royal Society presented Dalton with a Royal Medal,
he reserved his position as to the actual existence of atoms.[31] Davy
could not accept Dalton's postulate that the atoms of the various
elements were irreducibly different. He believed in the essential
unity of matter as a fundamental principle: Dalton's theory was
incompatible with this principle. Like Davy, Michael Faraday (1791–
1867) deplored the suggestion that there were a large number of
kinds of ultimate particle. He was disturbed by the discovery of
more and more metals and proposed that methods should be
devised to decompose them because he could not believe that
they were all elements. In fact, Faraday and Davy favoured the
notion of a non-material Boscovich-type atom, and Davy was
working on the problem of explaining chemical changes by means
of a Boscovich theory shortly before his death.

The problem of such large numbers of different atoms seemed resolvable by William Prout (1785–1850) who, in 1815, and anonymously at first, suggested that since atomic weights were near to whole numbers, the atoms of all elements might be composed of discrete numbers of hydrogen atoms. Prout developed his idea in his Bridgewater Treatise of 1834 and he related it to Ancient Greek theories of matter and suggested that hydrogen might be the primary basic element – the *protyle* – from which all other elements were formed. However, quantitative work done by other chemists, notably Jön Jacob Berzelius (1779–1848), showed that very many atomic weights were fractions. His work was so painstaking that the discrepancy could not be accounted for by suggesting that the analyses were inaccurate. Chlorine, in particular, had an atomic weight almost exactly between 35 and 36. There were various attempts to 'save' Prout's hypothesis, including William Crookes's (1832–1919) suggestion that atoms of the same element might differ in weight,[32] but there was no experimental support for this. The hypothesis was abandoned.

However, chemists did refer to atoms and atomic weights in spite of disliking what they held to be a metaphysical theory.[33] Berzelius published a table of atomic weights between 1818 and 1826 and adopted the formula H_2O for water having shown that Dalton's formula, HO, was incorrect.

EXPLANATION OF AFFINITY IN TERMS OF ELECTRIC CHARGES

Another weakness of Dalton's theory was that, like the corpuscular theories in the eighteenth century, it did little to explain *why* elements came together in particular ways in different compounds. Electric charge seemed to be the key to solving the problem of chemical affinity. It was known that solutions[34] of acids, bases and salts could be decomposed by passing an electric current through the solution. One set of elements and chemical groupings (cations) moved to the negative pole (the cathode, that is the negative electrode), the others (anions) going to the positive pole (the anode, that is the positive electrode). Cation elements or groups were electropositive, i.e. had a positive charge and were therefore attracted to the negative electrode; anion elements or groups were electronegative, i.e. had a negative charge and were attracted to the positive electrode. This was developed into a dualistic theory of chemical attraction which presupposes that matter is electrical in nature. Palmer quotes Gmelin:

> The atoms of each elementary body have two poles, on which the two opposite electricities are accumulated in different proportions, according to the nature of the bodies: for example, oxygen has a large negative quantity, and a small positive quantity, while potassium has the opposite disposition: hydrogen has nearly equal quantities of each electricity.[35]

Here was real progress, even though not all affinities could be explained in this manner. However, though an atomic theory is not incompatible with a dualist explanation of affinity, and this explanation is assumed by Gmelin, it is not a necessary presupposition.

REVIVAL OF DALTON'S THEORY

As a metaphysical speculation Dalton's theory was never entirely abandoned, indeed in an instrumental sense (see note 52) it was used by practising chemists. As Greenaway says, 'A profound dislike of metaphysical redundancy was turned aside time and time again by the expository power of Dalton's theory.'[36] Moreover, even those who were highly sceptical as to the physical existence of atoms appreciated the value of finding combining weights (equivalent weights) of elements and, therefore, whatever the view as to the reality of atoms, it was agreed that it was necessary to determine 'atomic' weights.

A SERIOUS ANOMALY AND ITS RESOLUTION

One serious objection to Dalton's theory that elements were composed of indivisible particles had arisen from studies of the combining volumes of gases. The ratios of the volumes were almost always in simple proportions and this had led Joseph-Louis Gay-Lussac (1778–1850) to propose the law that equal volumes of gases[37] must contain the same number of atoms. However, this would lead to atoms of certain elements having to divide in chemical reactions (for example, oxygen in the reaction of hydrogen gas with oxygen gas to give water): the theory seemed to entail that oxygen atoms would not be indivisible after all. Dalton himself refused to accept the suggestion and held that the law was a false generalisation based on dubious experimental evidence. His attempts to discredit Gay-Lussac's work by suggesting that his results were inaccurate were, quite rightly, unsuccessful. Many chemists, Berzelius for example, did accept Gay-Lussac's data, though they did not accept his explanatory hypothesis.

In fact, the anomaly was removed by Amadeo Avogadro (1776–1856) when he suggested that equal volumes of gases contained equal numbers of molecules. But that suggestion was rejected by Dalton because, from his work on the composition of the air (see above p. 61), he had concluded that atoms of the same element would repel each other. Dalton accepted that there might be molecules consisting of *different* atoms, but he thought that there could not be molecules composed of the same atoms. Another eminent chemist, Jean-Baptiste-André Dumas (1800–84), at first supported Dalton's theory, and Avogadro's, but his work on vapour densities[38] of gases, in 1837, made it appear that elements had different atomic weights at different temperatures. In 1841 it was shown that the number of atoms in a molecule might vary with temperature but, though this made varying vapour densities compatible with an atomic theory, it made that theory (along with Avogadro's) seem too complex to be useful. Avogadro's suggestion was ignored.

But in 1860 it was revived by Stanislao Cannizzaro (1826–1910). Cannizzaro did not convince his colleagues immediately, but eventually his exposition of Avogadro's explanation was accepted. It resolved earlier problems connected with the combining of volumes of gases and removed a serious objection to Dalton's theory. Atoms were indeed indivisible, but atoms in a molecule could separate. Further progress was to be made through the study of the effects of changes of temperature and pressure on gas expansion and contraction. A molecular theory of gases could account for many of the physical properties of gases.

MOLECULES

Corpuscular theories of matter were back in favour yet, even so, we find Clerk Maxwell (1831–79) writing in 1875, clearly aware of the explanatory power of a molecular theory but still not absolutely endorsing it as physical truth:

> Our definition of a molecule is purely dynamical. A molecule is that minute portion of a substance which moves about as a whole, so that its parts, if it has any, do not part company during the motion of agitation of the gas. The result of the kinetic theory,[39] therefore, is to give us information about the relative masses of molecules considered as moving bodies. The consistency of this information with the deduction of chemists from the phenomena of combination, greatly strengthens the evidence in favour of the actual existence and motion of gaseous molecules.[40]

Maxwell made the point that a notion of matter as a continuum would suffice for macro-investigation but would not be satisfactory when small quantities were being studied:

> Thus, if a railway contractor has to make a tunnel through a hill of gravel, and if one cubic yard of the gravel is so like another cubic yard that for the purposes of the contract they may be taken as equivalent, then, in estimating the work required to remove the gravel from the tunnel, he may, without fear of error, make his calculations as if the gravel were a continuous substance. But if a worm has to make his way through the gravel, it makes the greatest possible difference to him whether he tries to push right against a piece of gravel, or direct his course through one of the intervals between the pieces; to him, therefore, the gravel is by no means a homogeneous and continuous substance.
>
> In the same way, a theory that some particular substance, say water, is homogeneous and continuous may be a good working theory up to a certain point, but may fail when we come to deal with quantities so minute or so attenuated that their heterogeneity of structure comes into prominence. Whether this heterogeneity of structure is or is not consistent with homogeneity and continuity of substance is another question.[41]

In this article Maxwell described methods of measuring the diameter and mass of molecules. He admitted that results were no more than rough guesses, but he had no doubt that 'the determination of the mass of a molecule is a legitimate object of scientific research'.[42] He went on to discuss the limitations of early atomic theories and the Boscovichian centre of force theory, and he showed how subsequent nineteenth-century discoveries made these untenable. The early theories could not 'account for the vibrations of a molecule as revealed by the spectroscope',[43] and the Boscovich centres of force 'are no doubt of their own nature indivisible, but then they are also, singly, incapable of vibration'.[44] Thus the problems of an atomic theory of matter, noted by Leibniz and by Boscovich, remained, and there was the additional problem of accounting for vibration. 'An atom had to be able to behave as an undeformable body in its totally elastic collisions in a gas and yet vibrate in order to radiate, and that moreover in definite modes which could account for spectral lines*.'[45] Maxwell himself favoured a notion of the atom as a Helmholtz vortex ring*, with properties suggested by William Thomson,[46] later Lord Kelvin (1824–1907), though he appreciated that the theory carried enormous difficulties. However he suggested that these might be overcome.

The atomic theory seemed to survive some critical objections based on various analyses and measurements and it was to receive great support in that it could explain and relate the properties of *different* elements.

THE PERIODIC CLASSIFICATION

The atomic theory had never been rejected by all chemists and there had been attempts to classify the elements on the basis of atomic weights even before there was general agreement about the existence of atoms or the significance of atomic weights. Very soon after Dalton's theory was published, Johann Wolfgang Dobereiner (1780–1849) suggested a theory of triads; he pointed out that elements could be grouped in threes (for example, chlorine, bromine, iodine, and calcium, strontium, barium), which showed a gradual variation in chemical properties and also in atomic weights. But there were many exceptions to this rule; many elements could not be put into groups and some groups were bigger than three.

When the atomic theory was revived in the 1860s, John Alexander Reina Newlands (1837–98) formulated his Law of Octaves, pointing out that if elements were arranged in order of their atomic weights, then, with some slight adjustments,

> elements belonging to the same group usually appear on the same horizontal line.... the numbers of analogous elements generally differ by 7 or by some multiple of 7: in other words, members of the same group stand to one another in the same relation as the extremities of one or more octaves of music...[47]

But there did have to be some slight adjustment and the arithmetical series was only approximate. Therefore, many chemists still held that suggestions as to the actual existence of atoms were little more than fantasies. However, just a few years later, in 1869, when there was increasing confidence in the experimentally established atomic weights, and the anomaly apparently involved in Gay-Lussac's law had been removed by Cannizzaro, Lothar Meyer (1830–95) and Dimitry Ivanovich Mendeleef (1834–1907) – working independently – formulated a definitive law that the properties of elements varied in a periodic manner with their atomic weights.

Mendeleef produced his first table of the elements in 1869, making eight postulates:

1. The elements, if arranged according to their atomic weights, show an evident periodicity of properties.
2. Elements which are similar as regards their chemical properties have atomic weights which are either (or nearly) the same value (for example, platinum, iridium and osmium) or which increase regularly (for example, potassium, rubidium and caesium).
3. The arrangement of the elements, or of groups of elements, in the order of their atomic weights, corresponds with their valencies*.[48]
4. The elements which are most widely distributed in nature have small atomic weights and have sharply defined properties. They are therefore typical elements.
5. The magnitude of the atomic weight determines the character of an element.
6. The discovery of many yet unknown elements may be expected.
7. The atomic weight of an element may sometimes be corrected

by the aid of knowledge of the atomic weights of adjacent elements.
8. Certain characteristic properties of the elements can be foretold from their atomic weights.

It must be stressed that these postulates did not involve any metaphysical assumptions: 1–5 are empirical generalisations and 6–8 are predictions based on inductive inference. They were confirmed by direct observation and that indeed indicated that the periodic classification was useful. It also strongly suggested, even though no explanation was provided, that the classification might be based on something more than coincidence. Explanation was to come later, see chapter 6.

Mendeleef classed the elements into *series*, there being two short series of seven elements each, and then two long series of 17 elements each. The long series were themselves made up of two short series separated by a group of three elements, the *Mendeleef transition elements*. The elements were divided vertically into *groups*, and after the first two series the groups were divided into subgroups, A and B. Mendeleef showed that, in general, elements in the same group had the same valency[49] and that they showed a gradual variation in chemical properties, though the first members of a group were often rather (but not very) different from the others; for example, carbon, unlike other members of group IV, formed large molecules. As he had claimed, Mendeleef's classification enabled him to predict the properties of two elements unknown at the time, gallium and germanium, and also to check atomic weights of elements already identified, for example it was shown that iodine had a higher atomic weight than tellurium and therefore should be in group VII.

But there were anomalies: the position of cobalt and nickel (atomic weights having been checked) had to be inverted if properties rather than atomic weights were to be the criteria for membership of a group. Also the element hydrogen had no definite place and there was a good deal of controversy as to whether it should be classed with chlorine as a member of group VII, or as an alkali metal like sodium and a member of group I. For this reason there were still reservations by some chemists as to the existence of atoms.

ORGANIC CHEMISTRY AND THE ATOMIC THEORY

Animal and plant tissue is formed from compounds which have carbon as one of their constituents. Since compounds composed partly of carbon are characteristic of living organisms, they were originally called 'organic compounds'. Other compounds, the familiar mineral acids for example, were 'inorganic compounds'. When organic and inorganic compounds were first differentiated it was believed that the former could not be made in a laboratory. Indeed, it was surmised that in some way they held the secret of life and that they could only be formed by the tissues of living organisms. But in 1828 urea was made from an inorganic salt and this discovery showed that carbon compounds could, after all, be produced by man from non-living material. Thus started the process of experimental investigation of known organic compounds; it led not only to their production in the laboratory but also to the synthesis of new organic compounds not found in nature.

Such early experimental investigations into the composition and properties of organic compounds, both those made in the laboratory and those found in living matter, gave little support for Dalton's atomic theory; indeed the results of analysis seemed to militate strongly against it. For it soon became clear that many organic compounds were very complex. For example, it was found that there were a very large number of different hydrocarbons, that is compounds, consisting only of carbon and hydrogen, where the weights of hydrogen combining with a fixed weight of carbon were not in whole number ratios, i.e. Dalton's third postulate, the law of multiple proportions, was not obeyed (see above, p. 62 and also note 20). Also the formation of certain organic compounds made nonsense of Berzelius's dualist theory of chemical combination, a theory intimately involved with the atomic theory. For instance, Berzelius's theory could not account for the fact that in many organic compounds, hydrogen, which was taken as an electro-positive element, could yet replace chlorine, which was an electro-negative element. As we have seen (p. 72) the position of hydrogen posed problems for Mendeleef even though by 1870 the atomic theory was much more firmly established. Certainly, earlier in the mid-nineteenth century, organic chemists felt that they must set aside both Dalton's and Berzelius's theories; the general feeling was that the atomic theory was ingenious rather than correct.

VALENCY

Nevertheless, during the 1850s the idea of valency was developed. The notion of valency as the number of combining links in an atom whereby it can combine with links on other atoms is without meaning unless we accept that atoms do exist. This was shown to be especially helpful in developing some explanation of the properties of organic compounds, as revealed by chemical analysis. As the theory of valency developed it became apparent that, along with an atomic theory, it could explain the complex nature of organic compounds in terms of the arrangement of atoms in their molecules (the molecular structure). For example, hydrocarbons could be classified into different sets, each of which was characterised by a typical valency linking. In addition, differences in molecular structure could solve another problem: it had been shown that there could be several compounds, sometimes with very different properties, yet with the same composition by weight[50] – they were isomers*. Their different properties, and indeed the very existence of isomerism*, could be explained as being the result of different arrangements of atoms in the constituent molecules. The concept of large molecules built up from a 'skeleton' of chains or rings of carbon atoms proved to be invaluable in helping understanding of the enormous number of carbon compounds (both natural and synthetic) which were being isolated and produced.

There is no doubt that the molecular structure, based on a theory of atoms combining with fixed valency links, was essential to the progress of organic chemistry. This is confirmed if we consider the explanation of optical isomerism*.[51] In 1860, Louis Pasteur (1822–95) had suggested that the molecules of optical isomers might be mirror images, but the idea of valency links, or bonds, was still vague and his suggestion was not developed. In 1874 Joseph-Achille Le Bel (1847–1930) and Jacobus Henricus Van't Hoff (1852–1911) independently produced the theory that the valency bonds of carbon atoms were directed towards the four corners of a tetrahedron. This would entail that if the carbon atoms were joined to four different atoms (or groups of atoms) there could be two different molecules which would be mirror images; these molecules would be optical isomers.

Thus having at first appeared to undermine the atomic theory, organic chemistry eventually gave it strong support. For though it is not necessary to have an atomic theory to explain the constant composition of compounds or the laws of combining weights the theory presupposes such laws but the laws do not entail the theory (see also above, p. 61). Although it is not necessary to have an atomic theory to account for the experimental facts concerned with electrolytic decomposition, it is necessary to have an atomic theory to explain the occurrence and properties of isomers and the various forms of isomerism.

It remained the case, however, that some scientists – for example, Ernst Mach (1838–1916) – would only acknowledge an instrumental[52] interpretation of the atomic theory and thought it pointless to assume the actual existence of atoms. This is a positivist* view of knowledge and brings us back to the philosophical approach to the problem of matter. We must go back to consider the way that Berkeley and Hume treated that problem and relate their views to those of the positivists of the nineteenth and twentieth centuries.

SUMMARY

Chemistry became an organised science during the eighteenth century.

Even earlier it had been appreciated that chemical changes could be brought about by heat and it was generally thought that heat was an element. The Baconian theory of heat as motion was ignored by chemists. It was to be revived by Maxwell in the nineteenth century.

Theories of combustion were developed; the phlogiston theory was dominant and this did not depend on a corpuscular theory of matter. Many thought that hypotheses as to the nature of matter were unimportant and should be disregarded.

Towards the end of the eighteenth century the measurement of changes in weight in chemical reactions came to be seen as significant. Dalton published his atomic theory in 1805 and appealed to quantitative evidence from weighing for support.

His theory was welcomed at first but, though never entirely forgotten, it fell into disfavour. This was partly because it seemed incompatible with explanations of the behaviour of combining gases and also because it was not necessary to use it to account for chemical affinity and, in particular, with the dualistic theory of chemical attraction.

Later, the anomalies were removed and Dalton's theory was revived in the second half of the nineteenth century. It was shown that elements could be classified on the basis of their atomic weights.

Explanations of the properties of organic compounds on the basis of an atomic theory at first seemed to be impossible, but eventually the theory proved invaluable in the understanding of molecular structure.

Some scientists, influenced by positivism, were still reluctant to accept that atoms existed; they preferred to regard them as purely theoretical entities.

5

Phenomenalism: Matter as Sense Experience

In chapter 2 three different accounts of non-material reality were presented: Plato's Forms, Leibniz's monads and Boscovich's points of force. But though their accounts were all based on the view that ultimate reality was non-corporeal there was no denial of the existence of matter. Plato supported Empedocles's four-element theory, Leibniz held that bodies were well-founded phenomena, and Boscovich argued that the extension of matter was the result of repulsive forces. By contrast, the phenomenalist theories described in this chapter are based on the thesis that since our belief in the existence of corporeal bodies (and the matter or substance of which they are composed) depends solely on our having sense experiences, we should take any reference to bodies, to matter or to substance as ultimately signifying no more than a report of the possibility of, or of actually having, appropriate sense experiences of sight, touch, sound, taste and smell. Phenomenalists stipulate that any account of physical objects or bodies can be re-expressed as a description of actual or possible sense experiences and therefore physical objects should be regarded as nothing more than clusters of possible or actual sense experiences; terms referring to physical objects should, at best, be taken as a convenient name for the cluster. For phenomenalists there is no need to introduce notions of matter or substance as supporters of sensible qualities – they need no support.

Thus, though some phenomenalists may concede that physical objects might possibly exist,[1] they would stress that the question could not be settled and that therefore it would be pointless to discuss it. All phenomenalists agree that any reference to abstract matter or to substance as a support of sensory qualities is redundant.

Theirs is a purely philosophical approach to the problem of

matter and one that claims most emphatically to reject metaphysics. As indicated in the previous paragraph, phenomenalists support their view by inviting assessment of the empirical evidence we have for the existence of an external material world. That evidence consists solely of sense experiences. The concepts of matter and of physical objects are therefore irrelevant. These are, phenomenalists assert, merely the product of metaphysical speculation.

Despite the strict reliance on sense experience, phenomenalism is a philosophical, not a scientific, theory for it eschews interpretation and causal explanation. Nevertheless phenomenalists, in general, value empirical science. Scientific theories are not dismissed but they are regarded purely instrumentally[2] and not as giving descriptions of a reality behind appearances. Phenomenalists maintain that science cannot, any more than common sense, show what is 'really there' (if indeed anything is there). The function of science and value are in providing a means of prediction of future sense experiences and in enabling them to be ordered and classified. It is only in so far as it produces order and classification of experiences and potential experiences that science is explanatory.

BERKELEY

George Berkeley (1685–1753) has an anomalous position as compared with other phenomenalists because he explained the constancy and coherence of sense experiences which common sense takes as being caused by physical objects by reference to God. He may, therefore, be accused of introducing metaphysics by way of religious belief and, as we shall see, God had an important role in Berkeley's account of the nature of bodies. But it is still helpful to consider his views along with the more sceptical philosophers who followed him. What he has in common with later philosophers is the phenomenalist view that, at the last, talk of physical objects comes down to talk about sense experiences[3] and that references to substance or matter as a supporter of the qualities known through sense experience are misguided if not nonsensical. This is a form of idealism which Berkeley called 'immaterialism' because only immaterial sense experiences – Berkeley referred to them as *ideas* – are allowed to exist. But also, and somewhat paradoxically because of their reliance on sense, such philosophies can also be regarded as typifying a thorough-going empiricism.

THE ABOLITION OF SUBSTANCE

Locke had called sense experiences 'ideas of sensation' and he believed that they were *caused* by external corporeal bodies. As we saw in chapter 3, Locke held that the bodies we observed were composed of tiny corpuscles, too small to be sensed. The corpuscles had certain primary qualities and so did the bodies built from them, but bodies also had other, secondary, qualities because via their primary qualities, the constituent corpuscles had powers to produce sensations (ideas) in us. (See also chapter 3, p. 49). Locke thought that the primary qualities 'inhered' in a basic substratum, namely the *substance* of the corpuscles. He acknowledged that any such substance was unknown, and per-haps unknowable, but he thought that it must be there: to support the primary qualities:

> If anyone should be asked what is the subject wherein colour or weight inheres, he would have nothing to say but, the solid extended parts; and if he were demanded what is it that solid-ity and extension adhere in, he would not be in a much better case than the *Indian* . . . who, saying that the world was sup-ported by a great elephant, was asked what the elephant rested on, to which his answer was, a great tortoise; but again being pressed to know what gave support to the broad-backed tortoise, replied, something, he knew not what . . . The *idea then we have, to which we give the name substance, being nothing but the supposed, but unknown, support of those qualities we find existing.*[4]

BERKELEY'S IMMATERIALISM

Locke's view of substance is analogous to Aristotle's view of primary matter (*materia prima*) (see p. 15); neither substance nor primary matter possessed qualities, in effect they *supported* qual-ities. Berkeley went further than Locke; he suggested that if all we could know about substance and bodies came from sense experience, i.e. *ideas*, then we should accept that the *ideas* were all there was to know. It would follow that *all* observed qualities would be nothing but *ideas* before the mind and there would be no point in distinguishing primary qualities from secondary qualities. Moreover neither primary nor secondary qualities would exist without a perceiver, for *ideas* had to be in some perceiver's mind. In particular the collections of ideas commonly called physical objects existed only in the mind and there was no supporting substance. This is Berkeley's *immaterialism* – the thesis that *things* are nothing more than *ideas*:

> in this proposition, a die is hard, extended, and square; they [philosophers] will have it that the word *die* denots a subject or substance distinct from the hardness, extension, and figure, which are predicated of it, and in which they exist. This I cannot comprehend: . . . to say a die is hard, extended, and square, is not to attribute those qualities to a subject distinct from them and supporting them, but only an explication of the meaning of the word *die*.[5]

Although for Berkeley things were just collections of ideas, he did not reduce the material world to insubstantial images. Dr Johnson was said to have remarked that if he stubbed his toe on a stone he was well aware that there was more to that stone than an *idea*, but Johnson missed Berkeley's point. The hardness and sharpness of the stone were also *ideas* just as much as its visual appearance. Berkeley could further distinguish between *ideas* of the external world on the one hand, and images and dreams on the other: the latter were less vivid and less coherent. Moreover the *ideas* of what were thought of as physical objects existed independently of any mortal observer because they were God's ideas and depended only on Him. Human beings could perceive God's ideas and, as partly spiritual creatures, they could also have and perceive their own private *ideas*: they created their private and personal dreams and images. However, though each individual dreamer and imaginer could perceive his or her *ideas* whilst he or she was dreaming or imagining, they were unlike God's in that they were transitory and available only to the one individual creating them. By contrast, God's ideas could be perceived by everyone. In addition God *constantly* perceived His own *ideas* so that they were more permanent[6] than any human *ideas*. They had the permanence of what were customarily called physical objects.

> There once was a man who said 'God
> What seems most remarkably odd
> Is the fact that this tree
> Continues to be
> When there's no-one about in the quad.'

> Dear Sir,
> Your astonishment's odd;
> I am always about in the quad,
> And that's why the tree
> Will continue to be,
> Since observed by
> Yours faithfully,
> God.[7]

HUME'S SCEPTICISM

Hume also believed that we could only know sense experiences (*ideas*) but, unlike Berkeley, he was not prepared to invoke God as the creator of the special *ideas* which were constant and coherent. He acknowledged that we all took the existence of physical objects for granted, but he thought there was no point in asking whether physical objects really existed. Rather, we should inquire *why* we believed in their continued existence when not perceived and *why* we thought they were independent of our sense experiences:

> 'tis vain to ask, *Whether there be body or not?*
> . . .
>   We ought to examine . . . Why we attribute CONTINU'D exist-
> ence to objects, even when they are not present to the senses;
> and why we suppose them to have an existence DISTINCT
> from the mind and perception. . . . if the objects of our senses
> continue to exist, even when they are not perceiv'd, their
> existence is of course independent of and distinct from percep-
> tion; and *vice versa* . . .[8]

Hume argued that imagination prompted us to believe in the continued existence of our ideas:

> the imagination, when set into any train of thinking, is apt to continue, even when its object fails it, and like a galley put in motion by the oars, carries on its course without any new impulse. . . . The same principle makes us easily entertain this opinion of the continu'd existence of body. Objects have a certain coherence even as they appear to our senses; but this coherence is much greater and more uniform, if we suppose the objects to have a continu'd existence; and as the mind is once in the train of observing an uniformity among objects it naturally continues, till it renders the uniformity as compleat as possible.[9]

Hume was concerned to stress that there was no real ground for claiming that there are objects independent of perceptions; in particular there was absolutely no reason to claim that perceptions were *caused by* external bodies. He pointed out that our concept of causality and of one event causing another, arose from a regular and repeated experience of cause followed by effect. But, since we could *never* directly experience a body (physical object) since it was impossible for us to have any experience of a body followed by a perception. There were *only* perceptions:

> The only existences of which we are certain, are perceptions, which being immediately present to us by consciousness, command our strongest assent, and are the first foundation of all our conclusions. The only conclusion we can draw from the existence of one thing to that of another, is by means of the relation of cause and effect, which shews, that there is a connexion betwixt them, and that the existence of one is dependent on that of the other. The idea of this relation is deriv'd from past experience, by which we find, that, two beings are constantly conjoin'd together, and are always present at once to the mind. But as no beings are ever present to the mind but perceptions; it follows that we may observe a conjunction or a relation of cause and effect between different perceptions, but can never observe it between perceptions and objects.[10]

Hume concluded that perceptions were our only objects[11] and he also rejected and derided the hypothesis of the double existence of perceptions and objects, and the 'monstrous offspring'[12] of two principles, namely the interruption of perceptions and the continuance of objects. In his view we had no reason to believe that an external world existed; he thought that sceptical doubt 'is a malady, which can never be radically cur'd'.[13] Reliance solely on sense experience led Hume to solipsism* and scepticism; he maintained that only 'carelessness and inattention'[14] could afford a remedy. We shall see how close he comes to early twentieth-century attempts to rely solely on sense experience.

PHENOMENALISM AND POSITIVISM

Auguste Comte (1798–1857) put forward the thesis of positivism; he said it was founded on the scientific approach shown in the work and writings of philosophers and scientists such as Galileo, Descartes, Bacon and Hume. Comte claimed that positivist thinking, which he defined as reasoning from objective facts, would extend scientific methodology to all fields of human inquiry. There would be no contamination by metaphysics. In general, positivists were phenomenalists; they relied solely on sense experience and were not concerned with the ultimate nature of matter or the question as to whether physical objects existed. This was Mill's position; like Hume, he regarded such speculations as fruitless.

MILL'S PHENOMENALISM

John Stuart Mill (1806–73) was sympathetic to Hume's account and his reservations as to the existence of physical objects. He acknowledged that we had an intuitive belief in an underlying substratum, that is substance or matter,[15] but he was content to leave the problem of the ultimate nature of matter to the metaphysicians. He argued that whether or not there was some mysterious supporting substance, as far as we were concerned material bodies could be treated as collections of actual and possible sensations:

> Matter, then, may be defined as a 'permanent possibility' of sensation. If I am asked whether I believe in matter, I ask whether the questioner accepts the definition of it. If he does,

> I believe in matter.... In any other sense than this I do not.
> But I affirm with confidence that this conception of matter
> includes the whole meaning attached to it by the common
> world.... The reliance of mankind on the reality and perma-
> nence of possibilities of visual and tactile sensations.[16]

Perhaps Mill should have referred to sound, taste and smell for
these sense experiences can also constitute the cluster that makes
matter; but this is a side-issue because sense experiences of touch
and sight are undoubtedly what we primarily associate with matter.
It is interesting that Mill's comments are remarkably close to
Berkeley's account of the die. Both maintain that any reference
to matter or to physical objects is equivalent to reference to actual
or possible sense experiences. Berkeley called these *ideas*; later
philosophers would refer to *sense data*.

MACH'S PHENOMENALISM

Ernst Mach (1838–1914) had to accept that there was an involun-
tary belief in the existence of physical objects and in matter
and like Hume he suggested a psychological explanation for
this belief:

> In mentally separating a body from the changeable environ-
> ment in which it moves, what we really do is to extricate a
> group of sensations on which our thoughts are fastened and
> which is of relatively greater stability than the others, from
> the stream of all our sensations.... the sum of its constant
> elements as compared with the sum of its changeable ones,
> especially if we consider the continuous character of the tran-
> sition, is always so great that for the purpose in hand the former
> usually appear sufficient to determine the body's identity. But
> because we can separate from the group every single member
> without the body's ceasing to be for us the same, we are easily
> led to believe that after abstracting all the members, something
> additional would remain. It thus comes to pass that we form
> the notion of substance distinct from its attributes, of a thing-
> in-itself, whilst our sensations are regarded merely as symbols
> or indications of the properties of this thing-in-itself. But it would
> be much better to say that bodies or things are compendious

mental symbols for groups of sensations – symbols that do not exist outside of thought.[17]

Mach went further than Mill because he did not take a neutral position, he argued that the belief in the existence of physical objects (body) was definitely wrong:

> The crude notion of 'body' can no more stand the test of analysis than can the art of the Egyptians or that of our little children.[18]

In the extract below Mach repeated his contention that not only physical objects but also the entities postulated in scientific theories (for example, atoms) were nothing more than potential collections of sense impressions, and substance was nothing more than a mental construction:

> It is unquestionably very convenient always to have ready the name and thought for a group of properties wherever that group by any possibility can appear. But more than a compendious economical symbol for these phenomena that name and thought is not. It would be a mere empty word for one on whom it did not awaken a large group of well-ordered sense-impressions. And the same is true of the molecules and atoms into which the chemical element is still further analysed.
>
> True, it is customary to regard the conservation of weight, or, more precisely, the conservation of mass, as a direct proof of the constancy of matter. But this proof is dissolved, when we go to the bottom of it, into such a multitude of instrumental and intellectual operations, that in a sense it will be found to constitute simply an equation which our ideas in imitating facts have to satisfy. That obscure, mysterious lump which we involuntarily add in thought, we seek in vain outside the mind.[19]

It is worth bearing in mind that Mach was happy to include mass as a symbol in equations. It was a theoretical construct and did not signify any physical reality.[20] This can be compared to Heisenberg's view of atoms (see chapter 8, p. 136).

Mach contended that our physical world was simply the world of our sense experience and he called sense experiences the *elements* of the world. This is again a thorough-going idealism. Like Berkeley, Mach was asserting that *things* were nothing but *ideas*:

Let us look at the matter without bias. The world consists of colors [*sic*], sounds temperatures, pressures, spaces, times, and so forth, which now we shall not call sensation, nor phenomena, because in either term an arbitrary one-sided theory is embodied, but simply *elements*. The fixing of the flux of these elements, whether mediately or immediately, is the real object of physical research. As long as, neglecting our own body, we employ ourselves with the interdependence of those groups of elements which, including men and animals, make up *foreign* bodies, we are physicists.[21]

Mach wanted to establish some sort of objectivity so he distinguished the *elements*, which appear to have an objective quality, from individual sensations:

There exists . . . in the perspective field of every sense a portion which exercises on all the rest a different and more powerful influence than the rest upon one another. . . . In the light of this remark, we call *all* elements, in so far as we regard them as dependent on this special part (our body), *sensations*. That the world is our sensation, in this sense, cannot be questioned.[22]

But if *elements* are to be distinguished from sensations, it is difficult to understand how Mach thought we could ever know the former. The argument Hume used to show that body could not be said to cause perceptions (see p. 82) can surely be used to refute Mach. His conclusion that the world is our sensations is, effectively, that the world is *nothing more than our sensations*. Just as with Hume, if Mach's idealism (or empiricism) is pushed to its logical conclusion it reduces to solipsism.

OTHER PHENOMENALISTS

Karl Pearson (1857–1956) also asserted that bodies were nothing but sense experiences; he showed how paradoxes arising from atomic theories could be resolved if bodies (including atoms themselves) were not treated as objective realities:

the conception atom, when applied to our perceptions, is opposed to the conception of surface, as the continuous boundary of a body, we have here an important example of what is not an uncommon occurrence in science, namely two conceptions which cannot both correspond to realities in the perceptual world. Either perceptual bodies have continuous boundaries, and the atomic theory has no perceptual validity, or, conversely, bodies have an atomic structure, and geometrical surfaces are perceptually impossible. . . . But the whole difficulty really lies in the habit we have formed of considering bodies as objective realities unconditioned by our perceptive faculty. We cannot too often recall the fact that bodies are for us more or less permanent, more or less clearly defined groups of sense impressions . . .[23]

Émile Meyerson (1859–1933) did not affirm or deny the existence of an external world; nor did he assert that it existed. Like other phenomenalists he stressed that science did not attempt to give an account of ultimate reality, and also like them he sought to show how phenomenalism and scientific theories of matter were complementary:

Science . . . teaches us nothing about the noumenon; it only states precisely that there is a partial agreement between our intelligence and the external world. One may start from the fact of this agreement to arrive at the existence of the external world, as Leibniz did amongs others; but one may also use . . . the fact that this agreement is only partial, that there is also disagreement to prove the impossibility of the external world. . . . These are matters which are the exclusive property of metaphysics.[24]

A little later, Meyerson made the important point that there is no distinction between common sense and science in so far as everyday physical objects have the same status as the inferred entities of scientific theories. Both can be described as collections of sense data, 'there is an absolute continuity between science and common sense'.[25]

In effect, phenomenalists claim to abolish the problem of matter or substance (in the Aristotelian sense; see p. 15) by denying that there can be any significant discussion of something which, by definition, cannot be an object of sense experience. In addition, by maintaining that all we can know are sense data, they dismiss the question of whether or not there really are physical bodies as a topic for metaphysics. They are concerned only with phenomena, with the way the world appears. On that basis they claim that all statements about common-sense physical objects, as well as those about 'theoretical' scientific entities, can be 'translated', albeit perhaps rather long-windedly, into logically equivalent statements, or sets of statements about sense experiences, i.e. sense data. We now have to consider whether this is possible, i.e. whether in fact their purely philosophical solution of the problem of matter is successful.

THE WEAKNESS OF PHENOMENALISM

A.J. Ayer (1910–89) has offered cogent criticism. First of all he pointed out that the phenomenalist has to account for the possible existence of physical objects; it is not adequate merely to offer a psychological explanation for our belief that they exist. It is true that phenomenalists do not assert their existence, the question is left open (see below) but *if* there are physical objects, they must exist when not perceived – that is what we mean by 'physical object' (see note 1). Ayer makes this point when he distinguishes Berkeley's immaterialism from the phenomenalist position, *viz.*:

> He [Berkeley] did allow that things that commonly pass for physical objects could continue to exist when God perceived them: and to say that of something that it is perceived only by God is to say that it is not, in any ordinary sense, perceived at all.[26]

One might say that Berkeley avoided the problem of substance by converting it into a problem of the nature and existence of God.

As we have seen, phenomenalists contend that any statement about a physical object, or physical objects, can be 'translated' into a logically equivalent statement about sense data. Since, in general, phenomenalists have not directly denied the possibility that physical objects exist when not perceived, they have (with the exception of Berkeley) to complicate their account by bringing hypothetical statements into their 'translations'. It will be remembered that Mill wrote of the 'permanent possibility of sensation' (see above, p. 83). So, for example, 'There is a table in the next room' translates as '*If* anyone were to be in that room, *then* he or she would have the following visual and tactile sense data.'

However, as Ayer points out, there are very great difficulties for phenomenalists: first, if there is to be a specified observer or observers, they too must be described in terms of sense data. Even if one eliminates direct reference to a particular observer or observers, it is still necessary to describe the locality and the time – again solely in terms of sense data. These difficulties might be overcome but, as Ayer points out, the sense data statements must be such that it follows, logically, that the specified physical object, or objects, exist and this, Ayer thinks, cannot be done. Logical certainty can admit no conceivable doubt; it is not at all the same as empirical certainty:

> At the present moment there is indeed no doubt, so far as I am concerned, that this table, this piece of paper, this pen, this hand, and many other physical objects exist. I know that they exist and I know it on the basis of my sense-experiences. Even so it does not follow that the assertion of their existence, or that of the existence of any one of them is logically entailed by any description of my sense experiences. The fuller such a description is made, assuming all the evidence to be favourable, the more far-fetched becomes the hypothesis that the physical object does not in fact exist; the harder it is, in short, to explain the appearances away. But this is still not to say that the possibility of explaining them away is ever *logically* absent.[27]

Ayer goes on to show that it is even more doubtful that the existence of a physical object could be a logically sufficient condition for the occurrence of sense data.[28] He contends that when we refer to physical objects, we 'are elaborating a theory with respect to the evidence of our senses'.[29] Common-sense theories of physical objects and scientific theories explaining their behaviour and properties are empirically based on sense experience, but in so far as they transcend the evidence they have a metaphysical aspect.

Like positivists, phenomenalists hoped to be free of metaphysics and to banish all speculation beyond that based directly on sense experience. In doing so their ideal would be to reduce science itself to little more than classification – a natural history of sense experiences. But, if thrown out of the house, metaphysics has a tendency to re-enter through the back door. We have seen that an acceptable account even of common-sense physical objects (and also of scientific entities) must appeal to a theory which transcends the factual evidence. If we try to reject metaphysics and stick simply to basic sense experiences, we are left, as Mach shows in his attempt to distinguish *elements* from sensations, with either an arid idealism or a sterile solipsism.

But if we grant that there must be metaphysical assumptions embodied in our statements about physical objects and scientific entities, it is necessary to consider whether the problem of matter and substance can be dismissed as *merely* a topic for metaphysical speculation and of no interest to common sense or science.

Bertrand Russell (1872–1970) suggested that the speculations of Greek atomists were motivated by 'an instinctive belief that beneath all the changes of the sensible world there must be something permanent and unchanging'.[30] We saw the way the atomosists dealt with the problem in chapter 1. The belief in some underlying permanence is still with us; it has been successful in guiding scientific inquiry. The belief is tied to the concept of matter and that concept cannot, therefore, be abandoned. It must be reconstructed:

> the empirical successes of the conception of matter show that there must be some legitimate conception which fulfils roughly the same functions [as the older notions]. The time has hardly come when we can state precisely what this legitimate conception is, but we can see in a general way what it must be like . . .[31]

In our final chapter I hope to show how science and metaphysical speculation can come together to present an account of matter. But first we need to show how empirical theories about the nature of atoms were developed.

SUMMARY

Phenomenalists admit that physical objects may exist, but maintain that it is pointless to discuss the possibility; it is best to treat physical objects as clusters of sense experiences. They maintain that all references to physical objects are logically equivalent to references to actual or possible sense experiences. They wish to abolish appeals to the notion of substance as meaningless.

Berkeley's immaterialism (physical objects are clusters of *ideas*) is a form of phenomenalism, but Berkeley did allow for the permanancy of physical objects in that he maintained that *ideas* corresponding to what we call physical objects could be permanent since they were perceived by God.

Hume (1711–76) also thought that the only known existents were sense experiences and that it was 'vain' to discuss whether there were body or not. He admitted that his philosophy led to solipsism and scepticism, but saw no remedy.

There was strong support for the phenomenalists' position from positivism: the thesis that there should be no metaphysical speculation.

Mill, Mach and later phenomenalists accepted that there was an instinctive belief in the existence of physical objects but did not think it was justified. *Ideas*, the older term used by Berkeley and Mill, was replaced by *sense data*.

Ayer (1910–89) argued that it was impossible to show that any finite set of sentences describing actual and/or possible sense experiences can be logically equivalent to a statement describing a physical object or vice versa. He argued that reference to physical objects was justified by a theory based on the evidence of our senses. In so far as the theory transcends the evidence, it must have an irreducible metaphysical aspect.

6

The Divisible Atom

ATOMS AND ELECTRICITY

In chapter 4 (p. 66) it was stated that an electric current could decompose the solutes in certain aqueous solutions and that positive and negative ions moved to the cathode and anode respectively. This had led Sir Humphrey Davy to propose that chemical attraction was electrical in nature; a few years later Davy's idea was developed by Berzelius as the dualistic theory of chemical affinity described by Gmelin (see p. 67). Electricity was an exciting new phenomenon and the theory had, or seemed to have, wide philosophical implications. The German nature-philosophers of the late eighteenth and early nineteenth centuries had surmised that there was an ultimate and universal principle underlying phenomena and that electricity might provide an explanation of this principle.[1] They had speculated that chemical compounds were to be regarded as being built up from negative and positive parts and so were a unity of opposing entities.[2] Davy in particular was sympathetic to these ideas. Nevertheless, it was not philosophical speculation but the *experimental* work (of Davy and others), based on passing currents through various solutions, which gave strong support for a dualistic theory. As we saw in chapter 4, the suggestion was that in any compound there were two oppositely charged components held together by electrostatic attraction; the compound itself was electrically neutral.

But the theory had limited applicability even in relation to inorganic compounds and did not seem to apply at all to organic compounds (see p. 73). Also, since like electric charges (two positive or two negative) repel, the theory could not explain how two or more atoms of the same element, e.g. hydrogen, could combine together to form a molecule. It was therefore incompatible with Avogadro's hypothesis (see p. 68) in that Avogadro had proposed that those gases that were elements would be composed of molecules consisting of two or more of the same atoms. That hypothesis was ignored in 1811, but some 50 years later it was established; by then it was clear that a simple, dualistic theory of chemical affinity was inadequate.

On the other hand, it seemed certain that electricity played some part in chemical change and, for those who supported an atomic theory, it seemed that electric charges could reside on individual atoms. During 1833–4 Faraday had shown that a given quantity of electricity would decompose equivalent weights* of different elements and it was clear that there must be a very intimate connection between matter and electricity. However, Faraday himself was not disposed to support any atomic theory of matter; he thought that space was a plenum and rejected any notion of atoms and the void. Likewise he rejected any theory postulating discrete or ultimate 'atomic' units of electricity.

Despite such influential doubts, a few decades later Dalton's atomic theory of matter was gaining support and this revived and encouraged speculations as to the atomic nature of electricity. In 1871 Wilhelm Eduard Weber (1804–91) wrote:

> The relation of the two particles (of positive and negative electricity) as regards their motions is determined by the ratio of the masses e and e' on the assumption that in e and e' are included the masses of the ponderable atoms which are attached to the electrical atoms. [Note that here Weber leaves an open question as to whether the 'atoms' of electricity have weight.] Let e be the positive electrical particle. Let the negative be equal and exactly opposite and therefore be denoted by $-e$ (in place of e'), but let a ponderable atom be attracted to the latter so that its mass is thereby so greatly increased as to make the mass of the positive particle vanishingly small in comparison. The particles $-e$ may then be thought of as at rest and the particle $+e$ as in motion about the particle $-e$.[3]

As Palmer comments, if the signs of the particles are interchanged, Weber foreshadows later theories. He had presupposed an ultimate unit of electricity, though in 1871 this was still very much open to debate. Only two years later Maxwell wrote:

> for convenience in description we may call the constant molecular charge revealed by Faraday's experiments one molecule

of electricity, . . . it is extremely improbable that when we come to understand the true nature of electrolysis we shall retain any form of the theory of molecular charges.[4]

As things turned out, this was an infelicitous prophecy; change was rapid and the situation was fluid. The very next year, 1874, Johnstone Stoney (1826–1911) calculated the ratio of the positive unit of charge to mass on a hydrogen ion;[5] he called the unit of charge an *electrine*. That name fell into disuse, but later Stoney called both positively and negatively charged particles electrons.[6]

In 1881 Hermann Helmholtz (1821–94) wrote:

Now the most startling result of Faraday's laws is perhaps this: if we accept the hypothesis that the elementary substances are composed of atoms, we cannot avoid concluding that electricity also, positive as well as negative, is divided into portions which behave like atoms of electricity.[7]

It is interesting to note that investigations into the nature of atoms, and in particular their electrical properties, were being carried out by physicists rather than by chemists. In the last decade of the nineteenth century their work led to the gradual emergence of new ideas about electricity, namely that it was corpuscular in nature and that electrons embodied the ultimate unit of negative charge. These were new concepts which were only gradually accepted. Thus a few years earlier, in the 1880s, Joseph, John Thomson (1856–1940), who was to do so much in this field (see below), did not think that electricity was necessarily a material entity: it might be result of strain in the ether. He wrote:

The use of the word quantity (in defining the amount of electrification), . . . does not connote that electricity is a substance. . . . The method of expressing the state of electrification of a body, as due to the distribution of a certain quantity of electricity over it, in conjunction with older theories of electricity which supposed a quantity of electricity to be a quantity of fluid has caused electrical quantity to be looked upon as a fundamental conception in electricity. It has however no claim to be this.[8]

Later he came to the conclusion that Faraday's laws of electrolysis*
provided good evidence for there being ultimate units of electric
charge.

CATHODE RAYS

Cathode rays were first detected in 1858. It was discovered that
when a tube was nearly exhausted and a potential difference[9]
applied between two conducting plates at the two opposite ends,
a bright glow could be seen near the negative plate (the cathode).
Such glows were thought to be the result of some kind of radiation,
analogous to light, and, since they emanated from cathodes, they
were called *cathode rays*. But this was really just surmise because
the nature of the glow was unknown. In fact, in 1879, William Crookes
abandoned the notion that the glow was due to radiation and argued
that it might consist of a stream of particles. But the evidence was
still not clear and the term 'cathode rays' was retained.

It was shown that the rays travelled in straight lines, for an
object placed in their path cast a shadow. They also had a heating effect which could be demonstrated by focusing them with a
concave mirror (or concave anode). Neither of these observations
was incompatible with a radiation theory. However, in 1894
Thomson measured the velocity of the rays and showed that
though this was very high, it was nowhere near the velocity of
light. This counted against the radiation theory and in favour of
the suggestion that the effect was caused by streams of particles.
From 1895 to 1897 Thomson carried out further experiments and
showed that cathode rays were deflected by electric and magnetic fields. Here was yet more evidence for a particle theory.

RATIO OF CHARGE TO MASS: e/m

It was this last property (the deflection in electric and magnetic
fields) that provided Thomson with a means to discover, through
a series of careful and brilliantly designed experiments, the ratio
of the charge on each particle to its mass, i.e. e/m. It was found
that though the velocity of the particles varied with the potential difference across the tube, the ratio of charge to mass was
constant; it was also constant for all the different gases that were

put in the cathode ray tube. Later it was shown that negatively charged particles similar to those in cathode rays were given off by heated metal filaments, and they also had the same charge to mass ratio. A third method, beaming ultraviolet light on to a negatively charged plate, also produced negatively charged particles with the same charge to mass ratio.

A few years later Pieter Zeeman (1865–1943) was to show that atomic spectra[*10] were altered in magnetic fields. These spectra are due to rotation of electrons in their orbits round the nucleus, but in this context the relevant point is that the alteration in a magnetic field allowed a further calculation of the charge to mass ratio and gave the same result as the other three methods.

All these experiments tended to show that the electron did indeed represent a fundamental unit of electricity. Was that unit related to the electrically charged atoms (ions*) that could be found in solutions and also in electrically charged gases? The value of the charge to mass ratio of a hydrogen ion had been already determined earlier by Stoney (see above) and that value was some 1,800 times smaller than the charge to mass ratio for electrons. Three possible inferences were:

1. There was no relation between hydrogen ions and electrons.
2. Their masses were the same but the charge on the hydrogen ion was 1,800 times smaller.
3. Their charges were the same and the mass of the hydrogen ion was 1,800 times larger than that of an electron.

The third possibility was considered to be the most probable, and this was Thomson's own conclusion; he was convinced that his experiments showed that electrons carried a fundamental and ultimate unit of electricity. Thomson also experimented on various positive ions formed by charged gases and found that in these cases different gases had different values of charge to mass (e/m). This would be expected if the different ions had the same charge but different masses.

IMPLICATIONS FOR THE ATOMIC THEORY

Thomson's experimental work with electrons helped greatly to reassure chemists that atoms really existed, for even towards the end of the nineteenth century there were still some lingering doubts as to their nature; they might be centres of force or perhaps represent nothing more than a convenient explanatory concept. At the time, one particularly disturbing feature of the atomic theory was the discovery of more and more elements; this necessitated postulating the existence of more and more different kinds of atom. At the British association meeting in 1894 Lord Salisbury said:

> What the atom of each element is, whether it is a movement or a thing, or a vortex, or a point having inertia, whether the long list of elements is final, or whether any of them have a common origin, all these questions remain surrounded by a darkness as profound as ever.[11]

Thomson started his investigations the very next year and, by 1897, had established that there were 'free electrons' carrying an ultimate unit of charge. But many at first hesitated to postulate that electrons were *part of* atoms. According to Falconer,[12] how Thomson reached his corpuscular hypothesis must remain a matter of conjecture. There were speculations going back to Prout about atoms being composed of hydrogen units (see chapter 4, p. 66) but the distinguishing feature of Thomson's hypothesis was that his corpuscles were very small in size; electrons were not much more than 1/2000th of the weight of a hydrogen atom. Falconer suggests that Thomson was primarily interested in atomic *structure* and in electrons as part of atoms. She surmises that he was not much concerned with electrons as featuring in the propagation of electromagnetic effects:

> Thomson's interest in discharge, ever since 1882, had been focussed on atoms, their chemical nature and the bonds between them. The main attribute of his corpuscle was that it was a building block of atoms while, at least until 1900, the electron was the link between the ether[13] and a generalized matter. The corpuscle added to the electron concept its significance for atomic structure.[14]

Thus 'the suggestion that corpuscles were constituents of atoms was the most radical aspect of Thomson's theory.[15]

A REVOLUTIONARY PROPOSAL: SUB-ATOMIC PARTICLES

Thomson's experiments with cathode rays were carried out in the 1890s, and by 1900 he was arguing the case for electrons being a material part of atoms, strongly and without qualification. A necessary corollary of that view was that atoms were not, after all, indivisible: this was a revolutionary proposal. Many physicists objected because emission of electrons would seem to change the nature of an atom and introduce the possibility of transmutation. There was thought to be a danger in that such ideas might lead to a revival of medieval alchemy. However, there was so much experimental evidence that even those who had doubts began to entertain the possibility of sub-atomic particles. Thus George Francis FitzGerald (1851–1901), despite his reservations, was not absolutely opposed to Thomson's theory and even came to hope that Thomson might be right:

> In conclusion I may express a hope that Prof. J.J. Thomson is quite right in his by no means impossible hypothesis. It would be the beginning of great advances in science, and the results it would be likely to lead to in the near future might easily eclipse most of the other great discoveries of the nineteenth century.[16]

THE MASS OF ELECTRONS

At the beginning of the twentieth century only the *ratio* of electronic charge to mass had been established; for further progress some method had to be devised to find charge or mass separately. If one of these were to be measured, the other could, of course, be calculated from the ratio.

During the years 1913–18 Robert Andrews Millikan (1868–1953) devised a method of finding the charge on an electron. A very fine spray of oil was blown by an atomiser into a well-insulated chamber from where the fall of *very* small drops could be watched through a telescope. The velocity of fall of one particular drop was then measured. Next, the air (or other gas in the chamber) was ionised*[17] so that the drops, including the observed drop, acquired a small charge. After ionising, the velocity of fall altered, the alteration depending on the charge acquired by the drop; therefore, from the alteration, the charge could be calculated. That charge would vary, but it was always a multiple of a relatively small number which was taken to be *e*, the charge on an electron. From a series of similar experiments, all requiring very carefully controlled conditions, Millikan was able to demonstrate that the negative charge on an electron was equal and opposite to the positive charge on a hydrogen ion and so, as well as establishing the value, he confirmed the theory that *e* was a fundamental unit of charge.

ELECTRONS AND THE NUCLEUS

These experiments showing that the fundamental unit of electricity was the charge carried by an electron also showed that atoms were not the fundamental units of matter. It was discovered that electrons could be expelled from all atoms and so it came to be generally agreed that they must be discrete constituents of atoms. But atoms were electrically neutral and hence they must also contain something with a positive charge. Thomson had supposed that atoms might be spheres with a uniform positive charge spread over them, the electrons being embedded in the sphere rather like plums in a plum pudding. But the discovery of radioactivity and subsequent experimental work showed that this was incorrect.

RADIOACTIVITY

Radiation from uncharged materials was first detected by Antoine Henri Becquerel (1852–1909) in 1896. He was studying radiation discharges from ionised low-pressure gases. These discharges produced fluorescence, i.e. they could make crystalline materials glow. While investigating this effect Becquerel observed that a particularly penetrating and energetic discharge came spontaneously (and therefore without prior charging) from the element uranium. This is an element with a very high atomic weight. Marie Curie (1867–1934) and Pierre Curie (1859–1906) pursued this new phenomenon. They showed that during such spontaneous radiation, the constituent atoms decomposed. Thus the radiation produced was a *result* of decomposition, a radioactive decomposition. The Curies isolated radium and showed that it was one of a group of radioactive elements, all composed of heavy atoms. Their radiations could be differentiated by their power of penetration and the fact that the atoms of different elements took different times to decompose. Nevertheless all these *radioactive* elements gave out the same three types of radiation:

1. alpha particles, which Ernest Rutherford (1871–1937) showed to be relatively heavy, positively charged, ions;
2. beta rays, which were streams of electrons, i.e. negatively charged particles similar to cathode rays though moving somewhat faster;
3. uncharged gamma rays, which were like X-rays, though of shorter wave-length. They formed part of the spectrum of electromagnetic radiation (see p. 104).

ATOMIC STRUCTURE

Rutherford had established that alpha particles were helium nuclei with a double positive charge[18] and he made the surprising discovery that these relatively large and heavy particles were, in the main able to pass through thin foil of various metals. But about one alpha particle in every 20,000 was considerably deflected, sometimes so much that it doubled back on its course. In 1911 he explained this effect by suggesting that most of the mass of an atom was concentrated in a small and very dense, positively charged core or nucleus. He calculated that the diameter of the nucleus must be about 1/10,000th of the diameter of the whole atom. The nucleus was surrounded by negatively charged electrons which were at a relatively large distance from their nucleus. Rutherford surmised that they moved round the nucleus in orbits, rather like planets round the sun. These orbits were fixed and later came to be known as *shells*. Thus all atoms contained electrons of identical mass and (negative) charge; what *distinguished* the atoms of any given element from the atoms of any other element was the total positive charge on their nucleus.

ATOMIC NUMBER

Since atoms were electrically neutral, the charge on the nucleus had to be equal and opposite to that of the orbiting electrons.

The charge on each electron, *e*, was taken as the ultimate unit and the balancing positive charge on any atomic nucleus was *ne*, where *n* was the number of electrons. This number *n* was the *atomic number* of the element. It was atomic number rather than atomic weight which was characteristic of and unique to each element and it was atomic number which largely determined[19] its chemical properties and its place in the periodic table. Rutherford's model was supported by experimental evidence: heavier atoms deflected alpha particles to a greater extent than lighter atoms and this would be expected since they would have a more massive nucleus. Also it was discovered that emission of an alpha particle from a radioactive atom (and therefore a reduction of two in its atomic number) changed it to an atom of a new element, two places back in the periodic table.

NEUTRONS AND ATOMIC WEIGHT

The nucleus of the lightest atom, hydrogen, carried a positive charge which was, of course, equal and opposite to the charge on its single orbiting electron. A positively charged hydrogen nucleus was called a *proton*. It was found that with the exception of hydrogen, the atomic weight of most elements was about twice their atomic number. So the nucleus must consist of something more than just protons*.

During the second decade of the twentieth century a new method of determining the atomic weights of gases was developed. Thomson discovered that if electricity were passed through gases at low pressure, they were ionised, producing positively charged atoms. These could be deflected in electric and magnetic fields and so the ratio of charge to mass could be determined. Knowing the basic units of charge Thomson could calculate the mass of the gas ions and therefore their atomic weight. However he discovered that the same element might be composed of atoms of different weight. For example, most of the atoms of neon were mass 20 but a few (the ratio was 10:1) were of mass 22. Thus the atomic weight, as measured by chemical experiments, was 20.2.

At first this work was thought to show that each atomic nucleus was itself composed of protons and electrons. Thus, of the two isotopes of neon, one had a nucleus composed of 20 protons and 10 electrons, with another 10 electrons round it, whilst the other had a nucleus composed of 22 protons and 12 electrons, so still with 10 electrons outside the nucleus and therefore with the same atomic number. However, by 1920 Rutherford had shown that atomic nuclei were too small to be composed of so many particles and he proposed that they might contain particles of the same weight as protons but without the positive charge. These were called neutrons and by 1932 neutrons had been independently detected.

ISOTOPES*

It is atomic number that characterises an element. Atoms with the same atomic number but different atomic weight are called *isotopes*. *Chemically*, they are atoms of the same element because they have the same positive charge on their nuclei (and therefore the same number of orbiting electrons). But they do not have the same atomic weight because they have different numbers of neutrons on those nuclei. Here was an explanation of the exceptions to Prout's hypothesis (see Chapter 4, p. 66): it would seem that the atomic weight of each isotope was indeed a multiple of the atomic weight of hydrogen but chemical analysis yielded the *average* weight of the isotopes as the element's atomic weight. This average was unlikely to be an integer.

PACKING FRACTIONS

However, the story was not quite so simple. In 1927 Francis William Aston (1877–1945), who had been working with Thomson since 1910 and had been aware of the experiments on neon, showed that even the isotopes of elements might not quite be integers. By this time oxygen, with an atomic weight of 16, was defined as the standard so that, in theory, all isotopes should be multiples of one sixteenth of the weight of the standard oxygen isotope.[20] Aston found that very light and very heavy elements such as hydrogen and uranium had slightly higher atomic weights, whereas those of intermediate atomic weight, such as iron, might be of slightly lower atomic weights. Aston called the deviations *packing fractions* and the packing fraction was taken to indicate the stability of the atoms of a given element. This was especially interesting in regard to atoms of high atomic weight which were subject to spontaneous radioactive decay. It seemed that as they decomposed and formed more stable atoms they lost mass in the form of radioactive energy.

RADIANT ENERGY: ELECTROMAGNETIC RADIATION AND THE ELECTROMAGNETIC FIELD

We need to review the experimental and theoretical work on macro-effects in order to appreciate how radiant energy at the sub-atomic level was measured and how further discoveries were made. Years earlier, between 1856 and 1873, Maxwell had studied the electromagnetic effects produced by the relative motion of electrically and magnetically charged bodies and by electric currents. He had concluded that electromagnetic forces produced mechanical disturbances in the all-prevanding ether and that, in this way, they were able to travel through space as electromagnetic waves. On the basis of his observations he formulated equations to describe these waves. He showed that they travelled with a high but finite velocity, namely the velocity of light. Maxwell concluded that light itself was a mechanical disturbance in the ether which was set up by electromagnetic forces and that these forces could be regarded as establishing an electromagnetic field of force, analogous to the magnetic field of force revealed by iron filings scattered round a bar magnet. Bohr points out that Maxwell's theory of the existence of electromagnetic fields was the direct cause of the discovery of electromagnetic waves.[21]

THE COMPLETE SPECTRUM*

Maxwell's hypothesis that light was an electromagnetic disturbance of the ether was published as a treatise in 1873. It offered a synthesis of three physical phenomena: light, electricity and magnetism, but it did not win immediate acceptance; nor did it arouse any general interest. The significance of Maxwell's work became apparent only some ten years after his (relatively early) death as more and more kinds of electromagnetic radiation (that is electromagnetic waves) were detected. For example, Heinrich Hertz (1857–94), using Maxwell's equations, showed that radio waves were also electromagnetic waves, travelling at the same speed as light but of much longer wave length. Soon a whole spectrum of these waves from very short wave length – cosmic and gamma rays – through X-rays (see above p. 100), ultraviolet light, visible light, infra-red rays (radiant heat) and radio waves were detected and related. They were shown to form a complete spectrum, of which the spectrum of visible light (the colours of the rainbow) is only a small part.

SPECTRAL LINES

When elements are heated they may give out coloured light (for example, sodium gives out a bright yellow light and the colours of fireworks are due to other elements responding to heat) but they may also radiate energy in the non-visible part of the spectrum. In different circumstances they may absorb radiant energy and thus produce dark lines in the visible (and non-visible) spectrum. The radiation and absorption lines are characteristic of the atoms of each element. Atomic spectra were studied by Zeeman (see above, p. 96) and the macro-effects of electromagnetism led to theories of sub-atomic structure. As Bohr says, Maxwell had

> provided a rational basis for the wave theory of light . . . With the aid of the atomic theory it afforded a general description of the origin of light and of the phenomena taking place during the passage of light through matter. For this purpose, the atoms are supposed to be built up of electrical particles which can execute vibrations about positions of equilibrium. The free oscillations of the particles are the cause of the radiation, the composition of which we observe in the atomic spectra of the elements. . . . When the frequency of vibration of the incident waves approaches the frequency of one of the free oscillations of the atom, there results a resonance effect, by which the particles are thrown into specially strong forced vibrations. In this way a natural account was obtained of the phenomena of resonance radiation and the anomalous dispersion of a substance for light near one of its spectral lines.[22]

Refined methods for determining atomic weights had developed through spectroscopic analysis and this also helped to develop theories of atomic stability. Spectra provide evidence of processes by which an electron is added to an atom, its binding becoming more firm step by step with the emission of radiation. While the character of the binding of the other electrons remains the same, the steplike strengthening of the binding of this electron is visualised by orbits which at first are large compared with usual atomic dimensions, and become smaller and smaller until the normal state of the atom is reached.

[In X-ray spectra we witness] . . . the reorganization of the binding of the remaining electrons upon removal of one of the electrons previously bound. This circumstance, which has been especially emphasized by Kossel, was well suited for bringing to light new and important features of the stability of atomic structure.[23]

This was to throw new light on the nature of matter itself (see chapter 7).

ELECTRON SHELLS AND QUANTUM THEORY

The Rutherford atom model, of negative electrons circling a dense positive nucleus, was defective in that if electrons were continuously moving round the nucleus they should continuously emit radiant energy and they would, perhaps gradually but nevertheless inevitably, spiral into the nucleus. In 1913 Niels Bohr (1885–1962), working with Rutherford, showed how the model could be saved by appealing to the new quantum theory (see chapter 7). The theory postulated that energy, like matter, was atomic in nature and therefore radiant energy would not be emitted in a continuous band but only in discrete units. If electrons could circulate in stable orbits, or shells, energy would be emitted only when an electron 'jumped' from one shell to another. This would account for the fact that atomic spectra consisted of sharp lines rather than continuous bands since the energy of the quantum, and therefore the frequency of the radiation emitted when an electron 'jumped' from one orbit to another, would be sharply defined. Each line in a given atomic spectrum would indicate a definite quantum of energy as the electron went from one fixed (stable) orbit to another. The suggestion could also be related to the splitting of spectral lines in a magnetic field which had been noted by Zeeman.

In any shell no two electrons could have precisely the same energy and this meant that each shell could hold no more than a certain maximum number of electrons. The maximum increased with the shell's distance from the nucleus, shells being filled from those nearest to the nucleus outwards. A shell that had its maximum number of electron was closed and further electrons would occupy the next outer shell until it too was filled, and so on.

EXPLANATION OF CHEMICAL ACTION AND AFFINITY

During the 1890s chemically inert gases were isolated from the atmosphere. The spectral lines of one of them had been observed in light from the sun before it was detected on earth and it had been named helium. The other inert gases were neon, argon, xenon and radon – the last, radon, was radioactive. These gases had been difficult to detect and then isolate because they were completely unreactive chemically; they were also called the *rare gases*. It was suggested that they were so unreactive because their electronics shells were complete and especially stable. It was surmised that other elements would react so as to achieve a stable electron structure, like that of the rare gases.

ELECTROVALENCY

Affinity and valency could now be explained. If an atom has one lone electron in its outermost shell (the inner shells being filled) it can achieve greater stability by losing that electron, but of course it must be able to donate the particle to an atom prepared to receive it. Thus it will tend to react with atoms of other elements which are one electron short of a complete outer shell for the latter will achieve greater stability by receiving an electron. The atoms will react one to one so that both the elements would be univalent. Atoms of another element might need to achieve a stable structure by losing or gaining two electrons. They would be bivalent since one of their atoms would need to react with two atoms of the first kind. Bivalent atoms would indeed react one to one with each other, but the measure of valency is in relation to univalent atoms. Those atoms achieving greater stability by loss of electrons would form positive ions, whereas atoms achieving stability by gaining electrons would form negative ions. They would exemplify the dualistic theory of chemical affinity (see p. 67) and what came to be called *electrovalency*.

COVALENCY

But atoms with a larger number of electrons to gain, or lose, such as carbon atoms (which have four surplus electrons in their outer shell), would not form ions nearly so readily. This is because when one electron leaves an atom the positive ion so formed will hold the remaining electrons more firmly through electrostatic attraction; there will be an electrostatic force opposing the tendency to react to achieve a closed shell. This is why, in general, bivalent electrovalent elements are less reactive than univalent elements. Likewise, when an atom gains an electron and forms a negative ion there will be an electrostatic repulsive force opposing the tendency to achieve a closed shell by gaining yet another negative electron. Therefore, atoms with several electrons to gain, or lose, such as those of carbon, tend to share electrons in order to achieve a closed shell rather than parting with or receiving electrons (i.e. rather than forming ions). Two atoms of the same element as well as two atoms of different elements may share electrons to achieve the stable closed shell. The sharing of electrons exemplifies another kind of chemical bonding and is called *covalency*. Covalency is characteristic of organic compounds.

In 1915 Walter Kossel (1888–1956) wrote:

> Each successive element contains one electron and one elementary positive charge more than its predecessor. The fact that valency changes periodically proves at once that in passing from elements of lower to those of higher atomic weight the (electronic) configuration does not change uniformly. Instead, at regular stages, configurations are reached in which the number of electrons active in determining valency is repeated: configurations associated with chemical inertness also recur regularly; these are those of the rare gases. We conceive the property of valency to be essentially an aspect of the behaviour of the outermost electrons of an atom.[24]

Such a view of chemical affinity also explains the periodicity revealed by Mendeleef's table of elements: as the atomic number increases, so does the number of electrons. New shells are formed and with each new shell a new series of elements is produced. As they are arranged in the periodic table the elements in each series have the same number of electrons in the outer shell of their atoms and therefore similar chemical characteristics.

A successful account of chemical reactions and combinations had been formulated on the basis of atomic structure, but the nature of the sub-atomic particles, the constituents of atoms themselves, had to be elucidated. Their behaviour was so strange that they could not be adequately described as particles, that is as submicroscopic corpuscles. They seemed to have properties very different from the properties of macro-matter.

SUMMARY

Towards the end of the nineteenth century it became clear that there was an intimate connection between matter and electricity and that electricity must play some part in chemical change. Gradually it became accepted that atoms themselves existed and that electricity too was 'atomic' in that it consisted of discrete units of charge.

In the 1890s Thomson investigated the nature of cathode rays and showed that they consisted of streams of negatively charged particles, i.e. electrons. He estimated the ratio of their charge to their mass, e/m, and he argued that electrons were material parts of atoms, so initiating the revolutionary idea that atoms were divisible. Later Millikan devised a method for finding the charge on an electron and so both charge and mass were known.

Work with radioactive elements showed that atoms of heavy elements might disintegrate and Rutherford and Bohr suggested that atoms consisted of a positively charged relatively dense nucleus surrounded by the negatively charged electrons. The number of orbiting electrons gave the atomic number of the element and it was atomic number rather than atomic weight which characterised an element. Elements might consist of two or more atoms with different atomic weights but the same atomic number; such atoms were isotopes.

Further work showed that electrons were arranged round the nucleus in stable orbits or shells. This could account for the fact that each element showed characteristic spectral lines and could also explain the different types of valency shown by different elements.

7

The Duality of Matter

PARTICLES AND WAVES

The Bohr model of the atom (see chapter 6, p. 105) could account for chemical combination and it also helped in the interpretation of the structure of molecules of compounds. These new theories of molecular structure, particularly of the structure of organic compounds, explained the properties of many materials and could lead to the synthesis of entirely new compounds (for example, man-made products such as polythene) with very useful properties not found in 'natural' materials.

But the model posed problems for physicists, problems connected with atomic spectra such as those observed in the Zeeman effect. Bohr at this time (1913) presented a purely mechanical picture in which electrons were tiny charged particles orbiting around atomic nuclei. But his theory could not account for electrons jumping from one orbit to another (as demanded by the Zeeman effect); nor could Bohr explain why electrons did not steadily lose energy as they went round the nucleus. Later he wrote:

> The atom must have a stability which present features quite foreign to mechanical theory. Thus, the mechanical laws permit a continuous variation of the possible motions, which is entirely at variance with the definiteness of the properties of the elements. The difference between an atom and an electrodynamic model appears also when one considers the composition of the emitted radiation. For, in models of the sort considered, where the natural frequencies of motion vary continuously with the energy, the frequency of the radiation will change continuously during emission according to classical theory and will therefore show no similarity to the line spectra of the elements.[1]

In 1925 Werner Heisenberg (1901–76) suggested that these problems might be solved if mechanical models based on orbiting particles were be abandoned, and descriptions of atomic and subatomic events were to be based on the measured frequencies (and therefore wave lengths) of the observed spectral lines. In the same year Louis de Broglie (1892–1987) proposed something much more revolutionary, namely that electrons themselves should be regarded as having wave-like properties. Matter was to be intrinsically associated with electromagnetism and electromagnetic waves. Thus particles might sometimes be treated as waves.

It had already been shown that it was convenient to treat light rays as particulate packets of energy.[2] These came to be known as *photons* (see later, p. 121). De Broglie contended that a similar dualist approach to elementary particles would be fruitful, especially in explaining the properties of electrons. Indeed *any* material particle could be associated with a wave motion, though such association would be particularly helpful when dealing with subatomic particles. In 1926 Erwin Schrödinger (1887–1961) supported Heisenberg's explanation of the Zeeman effect and later it was shown that de Broglie's contention was correct and that not only electrons, but other sub-atomic particles such as protons, could be shown to have wave-like properties.

WHAT DOES DUALITY INVOLVE?

Above we have a brief sketch of events for the first three decades of the twentieth century, but in order to appreciate the new approach to the problem of matter and the full implications of its dual aspect we must consider the characteristics whereby we recognise waves as waves and solid particles as solid particles. This can best be understood by first reminding ourselves of the different properties of particles and waves at the macro-level. Some of these properties are very well known and are part of what might be called our common-sense knowledge of the world; they are so well known that they are taken for granted and are not explicitly formulated. For this very reason they may be only vaguely appreciated.

Other properties are less familiar, though they are well established by classical science and in fact evidence for their existence is based on common-sense interpretations of observation. In order to appreciate the fundamental reassessment involved in accepting the duality of matter the criteria whereby we ordinarily *distinguish* visible particles from visible waves need to be made explicit.

PROPERTIES OF MATTER

(a) Hardness and Impenetrability of Solids

We are all familiar with lumps of matter and the little lumps, such as grains of salt, which typify solid particles. They are hard and impenetrable and have a definite, perhaps a distinctive, shape.

(b) Weight and Mass

All matter, whether solid, liquid or gas has mass and in classical science it is more common to refer to mass rather than weight since the time that Newton had established that weight was the result of gravitational force acting on mass. Mass is therefore a more fundamental property of matter than weight – astronauts in space are weightless but still have mass. Nevertheless mass is clearly related to the familiar property of weight. All matter, from the largest rocks to grains of sand and on to atoms and sub-atomic particles, has mass (which may, at least in principle, be measured through weight) and it has inertia, so that a force is needed to change its velocity*.[3]

The Greek atomos theorists, the corpuscular theorists and Daltonian atom theorists thought of their ultimate particles as very tiny grains (tiny solid lumps) which were hard and impenetrable and perhaps had distinctive shapes.[4] When it was first shown that sub-atomic entities such as electrons and protons existed Thomson, Rutherford and Bohr conceived them as particles of this type. Heisenberg wrote:

> Atomic physics took as a starting point the apparently natural supposition that our knowledge of the atom will, with increasing accuracy of observation, perfect itself more and more. Though atoms represented the final indivisible 'brick' of matter, they nevertheless appeared to be miniature parts of ordinary matter. The atom then at least in our imagination, was endowed with all the macroscopic properties of matter.[5]

The ultimate constituents of matter were assumed to be particles: the smallest particles, the subatomic particles were organised in atoms, and atoms themselves might be joined in different ways to form molecules. Visible matter whether solid, liquid or gas was conceived as an assemblage of molecules.

PROPERTIES OF WAVES AND WAVE MOTION

We are also familiar with wave motion: a wave motion can be described as a disturbance travelling through a medium without the medium moving bodily in the direction of the waves. The disturbance is a form of oscillation in which the medium may either move backwards and forwards, oscillating in the direction the waves are travelling, or move up and down, perpendicular to the direction of the waves. The former is called *longitudinal* wave motion and is exemplified by sound waves. The later is called *transverse* wave motion and is exemplified by the more familiar waves travelling the surface of the sea or as ripples on the surface of a pond.

WAVE LENGTH AND FREQUENCY

Slightly more sophisticated concepts associated with waves (perhaps analogous to the concept of mass) are those of wave length and frequency. We do not often observe absolutely uniform or regular wave motion in the sea or in other water, but certain waves come near to this and so we have the concept of *wave length* – that is, the distance from the crest (or trough) of one wave to the crest (or trough) of the next wave, or from one point on a wave to the corresponding point on the succeeding wave. In a regular wave motion the number of waves in a given time is constant and is called the *frequency*.

So, from ordinary observation it is clear that macro-waves have very different properties from particles. First, particles may be at rest and then will have a fixed position, whereas waves are not at rest and have no fixed position. Second, particles have mass and are solid and impenetrable whereas waves do not have mass and are not solid and impenetrable. It is quite possible to swim through waves! To highlight the contrast, we might describe particles as objects and waves as events.

We now need to consider some special features of light waves and indeed of electromagnetic waves in general, since we have seen (p. 104) that light is only a part of a large spectrum of electromagnetic waves. These waves all travel at the same velocity, the velocity of light, and hence it follows that the frequency of the waves (that is the number of waves in a given unit of time) must increase as the wave length becomes shorter that is frequency of all electromagnetic waves is inversely proportional to their wave length.

LIGHT AND OTHER ELECTROMAGNETIC WAVES

A feature of *all* waves, and not just electromagnetic waves, is that, unlike particles, they cannot exist independently. Waves are essentially disturbances of a medium and there must be some medium for them to disturb. In the case of light there had been much debate but by the end of the nineteenth century it had become accepted that light, and electromagnetic radiation generally, consisted of waves travelling through a medium of ether.

The word 'ether' or 'aether' is Greek for the blue of the sky; in medieval times it was held to be the quintessence, the fifth element that Aristotle had said made the material of the heavenly bodies. In the seventeenth century the term had been appropriated by Descartes. He asserted that ether filled the apparently empty spaces of the heavens. For Descartes matter was logically equivalent to extension so that a vacuum was *logically* impossible and so-called empty space was a plenum suffused with ether. Philosophers and scientists were (understandably) vague as to the nature of ether, but it was generally thought to be a rather special sort of matter.

Robert Hooke (1635–1703) suggested that light might consist of vibrations of ether and not long afterwards Christian Huygens (1629–95) proposed that light rays might be regarded as waves in the ether. But his wave theory as to the nature of light was not generally accepted, largely because it was rejected by Newton. Newton favoured a corpuscular theory, a theory that light rays consisted of streams of tiny particles which were emitted from the source rather like bullets from a gun. He rejected the wave theory on empirical grounds because he thought that the observed behaviour of light, in particular refraction* and dispersion*, were much better explained by a corpuscular theory.

Newton contended that there were also theoretical reasons for denying that light was a form of wave motion. For, since from its definition any wave motion demands a medium through which the waves pass, it cannot occur in empty space and Newton thought it possible that heavens contained only the stars, planets and comets. For metaphysical and religious reasons he refused to accept Descartes' contention that a vacuum was logically impossible. In fact Newton did not reject the notion of ether entirely; and though he did not think it carried light waves, he conjectured that perhaps, as an empirical fact and not as a consequence of logical necessity, there might be some tenuous matter suffusing what appeared to be empty space. But he had little use for the concept and would indeed have been happy to reject it (see below).

Interest in the ether and in the wave theory of light revived as a result of the work of Thomas Young (1773–1829). Young modified and developed Huygens' suggestion that light rays were wave disturbances; he appreciated that there must be some medium in which the waves travelled and it seemed obvious that the ether must be that medium. Huygens had thought that light waves were like sound waves, that is *longitudinal* vibrations, and so he had pictured the ether oscillating backwards and forwards in the same direction as the direction of the light ray. At first Young favoured Huygens' view, but by 1817 he had come to the conclusion that light waves were more likely to be like water waves: *transverse* vibrations – the medium, that is the ether, oscillating up and down (vertically) whilst the waves themselves travelled forward (horizontally). Thus the ether was thought to vibrate at right angles to the direction of the light ray. Young's theory was ridiculed when he first proposed it (and was then ignored), but in a relatively short time there was a change and by 1825 the corpuscular theory of light was superseded. It was accepted that light rays were transverse waves in the ether.

For nineteenth-century scientists the luminiferous (light-bearing) ether was an established theoretical entity and had the same status as atoms and molecules. Like the ether, these too could not be directly observed and, as was stated in chapter 5, diehard positivists had rated them as nothing more than convenient notions for correlating observations. But most scientists had come to accept that though atoms and molecules might have special properties, they were as real as ordinary material objects such as tables and chairs. Likewise, although ether might have special properties it was as real a medium as water and air.

Indeed as the account given above makes plain, as far as late nineteenth-century scientists were concerned there *had to be* an ether. Its existence as a medium for the transverse wave motion was not a metaphysical presupposition related to belief in a plenum, but was a direct consequence of an empirical theory, Young's wave theory of light. If there were no ether there would be no way of explaining the transmission of light rays and the nature of light would have to be fundamentally reassessed. As further work showed that light was only part of a complete spectrum of electromagnetic radiation the wave theory seemed to be confirmed and the importance of the ether, as the medium of transmission, increased. Even those physicists who made no use of the ether save as a frame of reference* for electromagnetic equations did not want to concede that it had no real existence. For example, Hendrik Anton Lorentz (1853–1928) wrote:

> I cannot but regard the ether as endowed with a certain degree of substantiality, however different it may be from ordinary matter.[6]

Unfortunately the ether resisted all attempts to detect it.

In 1899 Albert Einstein (1879–1955), whilst not abandoning belief in the existence of the ether, pointed out that the concept could not be directly related to any observable physical effect. It was theoretically necessary, but since it could not be detected it had no intrinsic physical meaning; therefore it was superfluous *as a physical concept*:

> The introduction of the term 'ether' into theories of electricity leads to a notion of a medium of whose motion one can speak without, I believe, being able to associate any physical meaning with such a statement.[7]

This is remarkably similar to the view expressed by Newton nearly three centuries earlier:

> And as it [the ether] is of no use . . . so there is no evidence for its existence and therefore it ought to be rejected.[8]

FURTHER DISTINCTION BETWEEN PARTICLE MOTION AND WAVE MOTION

We have already considered basic common-sense distinctions between particles and waves. Two other characteristic features of waves, whereby they are distinguished from particles, may also be observed at the macro-level. They are *interference** and *diffraction**. Although these features are not as familiar as the differences previously discussed, they were well known to classical physicists long before the discovery of sub-atomic entities.

(1) Interference

(a) Particles
If we have a cross-fire of particles, some may pass without collision but those that do collide will suffer some permanent change in their direction, i.e. they will be deflected, and their velocity will be reduced; this will lead to loss of kinetic energy*. The alteration produced will depend on the relative masses and velocities of the particles. Thus there can be either no interaction or a very marked interaction. The situation in the first case might be described as being one of no interference; in the second case, where there is some permanent change, the term 'permanent interference' would have to be used.

(b) Waves
But if two waves cross paths, they will both continue unchanged after the cross-over. There will indeed be alteration of the medium in the region where the waves actually overlap because, *in that region*, but in that region only, the resultant disturbance will be the algebraic sum of the sets of waves. Thus, if two crests coincide, the resulting crest will be the sum of the two separate crests, but if a crest and a trough coincide the resulting disturbance will be less than either. If crest and trough are equal, the effect will exactly cancel out, leaving the medium undisturbed. *This* is interference and it is important to appreciate that interference is localised and occurs only in the region where the waves overlap. Elsewhere the waves behave independently.

Thus waves show interference whereas moving particles do not. We might say the contrast between particle and wave is the result of particles being impenetrable whereas travel through waves is possible.

(2) Diffraction

(a) Particles

We have seen that one reason why cathode rays were thought to consist of streams of particles travelling in straight lines was that an object placed in their path stopped the rays and therefore cast a shadow. This is in accord with common-sense experience: if a fusillade of bullets is fired at a barrier or screen, then, provided the barrier or screen is strong enough to resist the bombardment, further progress is stopped. If there is a hole in the barrier or screen which is larger than the bullets, then those that arrive at the hole will pass straight through and will strike a further screen placed on the far side. If there is a relatively large number of bullets, then those that pass through will give a clear 'outline' of the hole. There is no diffraction, the particles are either stopped by the barrier or continue on their path, not having been affected.[9]

(b) Waves

But if waves meet a barrier, a little bit of disturbance will creep sideways round the edges of the barrier so that there is no sharp outline. If there is a hole or slit in the barrier, and especially if this is small as compared with the wave length of the waves, it appears to act as a new source of wave motion. New waves 'fan out' on the far side of the hole. This is the effect known as *diffraction* and is another characteristic of wave motion.

Both interference and diffraction can be observed watching water waves. Two sets of ripples meeting in a pond will show interference in the area of water where the waves cross. Sea waves approaching a breakwater or a gap in a groin will show diffraction – the far side of the gap acts as a new source of the waves. One reason for Newton's concluding that light was *not* a form wave motion was that no diffraction was observed if light was passed through a slit; a sharp and clear bright patch was formed on a screen placed beyond the slit.

However, in the early nineteenth century Young explained the formation of rainbow-coloured bands often seen in soap bubbles as being due to interference between light rays reflected from the two sides of the thin film of the soap bubble. The colours of the rainbow are produced by light rays of slightly different wave length and since waves reflected from one side of the soap film would be slightly ahead (or behind) those reflected from the other side, the two sets of reflected rays would be out of step. Then there could be interference such that some waves (relating to one colour) would be enhanced, so that their particular colour showed bright, whereas others (relating to another colour) might be diminished or completely cancelled.

Newton had not observed diffraction because the slits he used were very large compared to the wave length of light rays; he had not appreciated that the wave length of light was so very, very short. But in the nineteenth century it was shown that when light was passed through a *very* narrow slit (with a width of the same order as its wave length) there was no sharp shadow. Then the emergent light ray was seen to 'fan out' and not to give a clear bright image; in other words, it showed a diffraction effect. This cannot be detected unless a very fine slit is used.

Thus it seemed that light had to be a form of wave motion and though there were worries about the nature of the medium, the ether, the wave theory of light was not seriously questioned. Moreover any corpuscular hypothesis would have been regarded as an *exclusive alternative* and, at the time a highly dubious alternative, to a wave theory. For, up to the end of the nineteenth century, there was no suggestion that any natural phenomena might be described in terms both of particles and of wave motion. Such a suggestion could come about only through a new appraisal of the interrelation of matter and energy far removed from the concepts of classical physics. That new appraisal was forced on scientists as they further investigated radiant and electromagnetic energy and as they continued their experiments with sub-atomic matter.

THE PHOTOELECTRIC EFFECT

At the beginning of this chapter it was stated that the concept of photons introduced the notion that in some circumstances light waves might be treated as having a particulate aspect. The concept emerged as a result of study of photo-electric effects. Right at the end of the nineteenth century, Philip Lenard (1862–1947) had shown that when light is shone on to a metal plate, electrons are ejected from its surface. This is the *photoelectric effect*. Lenard showed that the light had to be of a certain threshold frequency, f_o to produce any effect at all but, above that threshold the energy, E, of the emitted electrons was proportional to the difference between the frequency of the shining light f and the threshold frequency f_o, That is,

$$E \text{ was proportional to } f - f_o$$

and did *not* depend on the intensity of the light. It seemed strange that the intensity of the light did not have any influence on the energy of the electrons. The explanation involved appeal to Planck's new theory as to the nature of energy.

QUANTUM THEORY: A NEW APPROACH TO ENERGY

In 1900 Max Planck (1858–1947) had studied the radiant heat given out by black bodies* at various temperatures and had shown that the relationship between the temperature of the body and the heat radiated could be explained only if the rays were emitted not as a smooth continuum, but rather in very small but discontinuous pulses. The discontinuity was the result of there being indivisible units of action*,[10] which he called *quanta*. This was Planck's Quantum Theory. The central feature was that because the quantum of action was indivisible, it followed that action, and in effect energy (the product of action and frequency), could not be divided into indefinitely smaller amounts. Thus, if the theory were accepted, energy, like matter, must be atomic and consist of discrete 'packets', the size of the packets being determined by the ultimate quantum. The quantum is taken as a physical constant and is called Planck's constant. It can be measured experimentally and is denoted by h.[11]

LIGHT WAVES AS PARTICLES

In the early twentieth century Einstein pointed out that though the movement and speed of light could be explained in terms of a wave theory, that theory could not explain how light interacts with matter. To do this, he said, it would become necessary to assume that when light hits a surface it behaves, not as a set of rays of continuous energy, but (rather like the radiant heat studied by Planck) as though it were made up of small packets of energy. These are the *photons* (see p. 111). Indeed, Einstein appealed directly to Planck's quantum theory, stating that the energy of a photon must be given by hf where h is Planck's constant and f is the frequency of the light. Then it would follow that the energy, E, of the emitted electrons (see above) would be given by

$$E = h \left(f - f_0 \right)$$

If the intensity of the light increases, the *number* of emitted electrons increases but their energy remains the same.

Einstein said:

> The usual conception that the energy of light is continuously distributed over the space through which it propagates encounters very serious difficulties when one attempts to explain the photoelectric phenomena as has been pointed out in Herr Lenard's pioneering paper.
>
> According to the concept that the incident light consists of energy quanta, however one can conceive of the ejection of electrons by light in the following way. Energy quanta penetrate the surface layer of the body and this energy is transformed, at least in part, into kinetic energy of electrons.[12]

At that time, a return to a corpuscular theory of light was revolutionary and Einstein made clear that the wave theory was still applicable in many contexts. It was just that a corpuscular theory could best explain how light interacted with matter. He wrote:

> The wave theory of light, which operates with continuous spatial functions, has worked well in the representation of purely optical phenomena and will probably never be replaced by another theory. . . . In spite of the complete experimental confirmation of the theory as applied to diffraction, reflection, refraction, dispersion*, etc., it is still conceivable that the theory of light which operates with continuous spatial functions may lead to contradictions with experience when it is applied to the phenomena of emission and transformation of light. . . . In accordance with the assumption to be considered here, the energy of a light ray spreading out from a point source is not

continuously distributed over an increasing space but consists of a finite number of energy quanta which are localised at points in space, which move without dividing, and which can only be produced and absorbed as complete units.[13]

Heisenberg points out how this implication of a corpuscular aspect to heat rays showed an inherent contradiction in the classical concepts of wave and particle:

After the discovery of the quantum of action by Planck the first and most important step was the recognition (achieved by Lenard's investigations and their interpretation by Einstein) that light, in spite of its wave nature as shown by countless experiments of interference, nevertheless does show corpuscular properties in certain experiments. Thus again we find classical physics, at the beginning of the new theory, involved in inner contradiction when attempting an interpretation of certain experiments entirely consistent with its own principles. In Bohr's atomic theory, which was based on Rutherford's experiments, the dualism alien to classical and earlier physics came even more clearly to the fore.[14]

Later work with X-rays showed that all electromagnetic radiation had a dual aspect.

X-RAYS AS PARTICLES WITH MOMENTUM

In 1923, Arthur Holly Compton (1892–1962) discovered that the wave frequency of X-rays was reduced when they collided with and were scattered by electrons, that is to say they lost energy. The scattering and loss of energy is called the Compton effect. It is analogous to the change in direction (deflection) and reduction in speed (therefore reduction in kinetic energy) which would be expected if a moving body collided with another body. Compton also observed that the reduction in frequency, and therefore the energy loss, increased with the angle of scattering. Likewise, on collision material bodies lost more energy and momentum as the collision angle of deflection increased. It followed that in these circumstances X-rays could be regarded as small moving material bodies, i.e. moving particles.

Moving material bodies have a momentum*, p, which in classical physics is the product of mass and velocity. Compton showed that the momentum of the X-rays could be expressed as:

$$p = h/l$$

Momentum equals Planck's Constant divided by Wave Length, l.

As stated above, as the wave length of electromagnetic radiations increases the frequency is reduced, so that an increase in wave length entails a reduction in energy and a reduction in momentum. The equation above shows this: for as l increases, the fraction h/l, that is the momentum, becomes less. The electron particles scatter the X-rays and their energy is reduced just as if those rays were themselves particles. We should note that the energy loss of the X-rays is measured as a reduction in *wave frequency*, not as a reduction in velocity (as it would be for an ordinary material particle). There is a duality here for *in these circumstances* the X-rays behave like particles but they still show some characteristics of waves.

PARTICLES AS WAVES

Lenard's experiments, Einstein's analysis and Compton's work established that in certain circumstances it was more helpful to treat these electromagnetic waves as particles. But what about the standard material particles? As was mentioned at the start of this chapter, de Broglie proposed that particles might be treated as waves in relation to solving problems concerned with atomic structure. He contended that just as light rays could behave like particles, a similar dualism could be applied to particles – they could have wave-like properties. He developed a theory in which any moving particle could be associated with a wave motion. The wavelength, l, would be related to Planck's constant and the momentum, p of the particle, and would be given as:

$$l = h/p$$

that is, Wave Length equals Planck's Constant divided by
Momentum, p.

De Broglie suggested that, considered as a form of wave motion, an electron would have to have an orbit round the nucleus which

was an integral number of wave lengths. Then, as result of interference, the electron/wave would superimpose at each 'circuit'
and so would be apparently motionless and stable. The stability
was essential, and only those orbits that were an integral number
of wave lengths would be possible.

ELECTRONS AS WAVES SHOWING DIFFRACTION

But even though, for explanatory purposes connected with atomic
structure and spectra, electrons might be conceived as behaving
like waves as well as being particles, can we show that they may
behave like waves in other circumstances? We saw above that
one feature which distinguished particles from waves at the macro-
level was that particles do not show diffraction effects whereas
waves do. So, if electrons always behaved as particles, they would
pass through a slit in a screen showing a sharp 'outline' of the
slit; they would show no wave-like 'spread'. But if de Broglie's
theory is correct and electrons can behave like waves, they will
'fan out' beyond the slit, i.e. they will show diffraction. This is
all very well in theory, but demonstration is difficult because diffraction effects will be observable only if the slit is extremely
narrow. Why is this?

It is important to appreciate that since Planck's constant is very
very small (see note 11) the wave length associated with even a
tiny particle like an electron (which has a tiny mass and therefore a tiny momentum) is also very very small, much less than
the wave length of visible light. That is why, if we are to demonstrate diffraction effects with electrons passing through slits,
we must have an extremely narrow slit. (It will be remembered
that the slit must be as small as the wave length of an approaching wave and it was because the slits he used were too wide to
show a diffraction effect that Newton did not think that light
could be a form of wave motion.)

The narrowest slits we can construct are all much too wide for
electron diffraction to be observable but it is possible to find narrow
enough slits by using the gaps between crystal atoms. In a crystal the atoms are arranged in a regular, three-dimensional pattern
and the space between two regular rows of atoms can serve as a
fine slit. Photographs of the paths of electrons transmitted through
thin crystalline foils do show diffraction and also interference

patterns, like those seen with water waves but on a much much smaller scale. Nevertheless scale is immaterial; by the end of the first quarter of the twentieth century it was established that electromagnetic waves could behave like particles and that particles could behave like waves. Heisenberg highlights this problem:

> It was the increased range of technical experience which first forced us to leave the limits of classical concepts. These concepts no longer fitted nature as we had come to know it. We observed the track of an electron moving as a particle ... and, on another occasion we found it reflected on a diffraction grating like a wave. The language of classical physics was no longer capable of expressing these two observations as effects of a single entity. We had, first of all, to define more closely those places where classical concepts became ambiguous in their application.[15]

Did this mean that all classical concepts of matter were fundamentally flawed, or was there some hope of reconciling the two approaches, some way of showing that the two accounts described different aspects of some more fundamental reality? One indication that reconciliation might be possible was that the two equations relating momentum and wave length, as given by Compton and de Broglie, connect those same entities, wave length and momentum, through Planck's constant:

$$Compton \ p = h/l$$

Here we have the relation whereby electromagnetic waves may be treated as moving particles and therefore may be assigned a momentum.
And:

$$de \ Broglie \ l = h/p$$

Here we have the relation whereby moving (orbiting) particles may be treated as waves and therefore may be assigned a wave length.

Thus we can say that electromagnetic waves have particle-like properties and sub-atomic particles have wave-like properties. This necessitates removing the common-sense and classical science distinctions between particles and waves. It must be stressed that the *particle aspect* of electromagnetic waves is shown more clearly with high-frequency, high energy rays such as X-rays. Compton's equation ($p = h/l$) tells us that their momentum is inversely proportional to their wave length and therefore directly proportional to their frequency. By contrast the *wave aspect* of particles shows more clearly with low-mass particles (assuming the same velocity) since de Broglie's equation ($l = h/p$) tells us that their wave length is inversely proportional to their momentum, and therefore (for any given velocity) inversely proportional to their mass. This is what we would expect: the particle aspect of waves becoming more apparent with waves of very short wave length and the wave aspect of particles becoming more apparent with particles of very small mass. The wave aspect of macro-particles and the particle aspect of longer waves is far too insignificant to detect.

Nevertheless it is there and entails a new concept of matter based on its dual nature. Can there be any way of reconciling the new properties with our familar concepts of particles and waves? In 1920 Erwin Schrödinger (1887–1961) suggested that our ideas of the properties of particles and of waves could be related through the concept of the wave packet.

WAVE PACKETS

One obvious distinction between waves and particles seems to be that waves are intrinsically associated with motion and therefore have no fixed position, whereas particles, even when moving, have a determinable position at a particular place *at any given time*. Schrödinger proposed that these two characteristics could be made compatible by treating them as at least made partly complementary in the concept of a wave packet.

We think of waves as consisting of a series of crests and troughs appearing in some medium, and if the wave motion is regular, the crests and troughs are of equal height (that is, they are of the same amplitude) and equally spaced (that is the wave length is constant). The medium itself vibrates but stays essentially in the same place though the waves travel onwards. Typical examples already instanced are sea waves and ripples in water caused by throwing a stone into a pond. Another example may be produced by jerking a piece of rope up and down to produce a set of waves.

Here, then, we have our familiar examples of wave motion. In these cases an approximate wave length is apparent but the waves have no fixed position.

It would be difficult to produce just one water ripple from a stone thrown into a pond, but it is possible to produce a single travelling wave in a piece of rope by jerking it just once. This one wave would travel along the rope. Since there is only one crest, we cannot speak of a wave-length but at any given time we can give the position of the wave. Such a single wave is called a *wave packet*. Of course, a wave packet in a rope would have a transitory existence.

ONE WAVE PACKET FROM SEVERAL WAVES

But we can produce a more stable wave packet by making use of the wave property of interference. As we saw (p. 117) two overlapping waves produce a resultant wave which is the algebraic sum of the component waves. In the region of overlap there would be no sign of disturbance if the waves were to cancel each other; on the other hand, there might be a double disturbance or there might be something in between. This occurs with two waves but the same algebraic effect is produced by allowing many different waves to interfere and it is possible to calculate mathematically what the effect of superimposing different waves would be. If we take just two different waves the regular motion of either one separately is broken up into a set of wave packets, but as more and more different waves are added one particular wave packet can be built up whilst the others are reduced. If we select wave components appropriately, one single crest (the wave packet) can be produced whilst elsewhere there is no disturbance in the medium.

A wave packet is built from a multitude of superimposed waves and so exemplifies wave motion, but it differs from familiar wave motion in that it has a definite position and no wave length. This, of course, shows that it has particle-like properties. The wave packet is to be contrasted with a simple wave motion of common sense and classical physics where there is a definite wave length but no definite position. These are two extremes; in between there are sets of wave packets with approximate wave lengths and approximate positions.

This is how Schrödinger's account of wave packets can bridge the distinction between waves and particles for wave packets have both wave-like and particle-like properties. However we must bear in mind that if wave length cannot be measured, then the frequency (which is inversely proportional to wave length) cannot be measured and therefore the energy of the wave packet is not known. In addition, even a clear-cut wave packet has a little 'spread' and hence its position cannot be precisely measured. Are there corresponding difficulties for particles?

In the next chapter we shall see that the two aspects of matter can also be reconciled by the assimilation of matter and energy. We shall also consider what seems to be an inherent uncertainty in ascertaining energy and position which is analogous to the uncertainty inherent in a wave packet.

SUMMARY

Through the study of atomic spectra and of the photoelectric effect it became apparent that, at the sub-atomic level, matter might best be treated as having wave-like properties and that in certain circumstances electromagnetic waves behaved like particles.

The atomist's notion of matter being constituted of tiny hard and impenetrable particles had to be abandoned, as did the notion of electromagnetic radiations consisting of waves propagated through the ether.

Sub-atomic particles had a wave length associated with them and could show interference and diffraction, and electromagnetic waves had a momentum associated with their frequency.

These two sets of properties were related through the quantum of action, Planck's constant. De Broglie had shown that the wave length of any particle of momentum p was equal to Planck's constant divided by p, and Compton had shown that the momentum of any electromagnetic wave of wave length l, was equal to Planck's constant divided by l.

Thus matter had a dual nature: it could be described in terms of particles or in terms of waves. The apparent incompatibility of properties could be reconciled in Schrödinger's concept of the wave packet.

8

Matter and Energy: Abstractions and Probability

ENERGY AND THE MOTION OF MATTER

Until the early twentieth century scientists studied the structure of matter and the movement of matter as two separate subjects. All matter might possess energy in virtue of its position and it would always possess energy if it were moving. Thus matter in motion was intrinsically connected with energy, but matter was not the same as energy; the two were distinct. Matter was particulate and essentially inert, whereas energy was immaterial and essentially active. Energy was evinced without matter in electromagnetic radiation.

ENERGY AND ELECTROMAGNETIC WAVES

For example, it was known that heat was a form of energy and, in material bodies, the movement of the component molecules produced heat and so would raise the temperature of the body. But the heat of infra-red rays (and indeed visible light) was a form of immaterial energy; all electromagnetic radiation exemplified an energy which seemed independent of matter.

It was stated in chapter 7 that the energy of electromagnetic waves depends on their frequency and, since we know that their frequency is inversely proportional to their wave length, the longer their wave length the less their energy. Thus wireless waves, with a relatively very long wave length[1] have much less energy than visible light and the latter has less energy than X-rays. High frequency high energy waves are more penetrating and therefore more dangerous. For example X-rays can penetrate flesh whereas visible light cannot; in addition, and because of their high energy, they are potentially harmful.

IMMATERIAL REALITY

(a) Spiritual

Some philosophers have suggested that, fundamentally, reality was immaterial (see the discussion in chapter 2). However, they did not think that the common-sense belief in the existence of a material world was simply false and that our senses deceived us. Plato did not deny the existence of matter even though he thought that the immaterial Universals, accessible to the mind but not to the senses, were the true reality. Leibniz held that reason showed metaphysical reality to consist of immaterial souls, but nevertheless he held that visible and tangible matter had phenomenal reality. In chapter 5 we saw that Berkeley believed that material entities were ideas conceived and perceived in the mind of God and that they had permanent existence as long as God continued to perceive them. These philosophers did not deny that matter had a phenomenal existence but they insisted that ultimate reality had an essentially spiritual aspect.

(b) Physical

If we grant with Boscovich (see p. 28) that atoms should be regarded not as particles but as points of force, then we have another route for resolving the problem of the duality of matter; it can be resolved in the concept of matter as energy.

Atoms, originally conceived as material particles, become one form of energy and wave motion becomes another form of energy. This is more than a merely philosophical solution a reassessment of concepts, because there is experimental evidence and also theoretical justification for such a view. Mass can change spontaneously into energy, and it can also be converted into energy. Aston's discovery of packing fractions (see Chapter 6, p. 102) showed that unstable radioactive atoms lost mass as they emitted electromagnetic rays. Einstein's work entailed that mass was a form of energy.[2] Since electromagnetic radiation is a form of matter, 'matter and radiation are different effects of a unified (einheitlich) event'.[3] The contrast, made in chapter 6, whereby matter (particles) and an electromagnetic energy (waves) were differentiated: matter as object and wave as event, has disappeared. Matter and waves may both be regarded as events.

ATOMS AND SUBATOMIC PARTICLES AS ABSTRACTIONS

It follows that the classical concept of atoms as particles, along with the neoclassical concept of subatomic particles, has to be abandoned. As Heisenberg writes, they 'are no longer material bodies in the proper sense of the word':

> Modern atomic theory is thus essentially different from that of antiquity in that it no longer allows any reinterpretation or elaboration to make it fit into a naïve materialistic concept of the universe. For atoms are no longer material bodies in the proper sense of the word, and we are probably justified in claiming that in this respect modern atomic theory embodies the principal and basic idea of atomic theory in a purer form than did ancient theory.[4]

But even if the tiny impenetrable material particle is no longer thought to exist, it can still have a symbolic use. Heisenberg compares it to the square root of -1. In fact there is no such square root yet even so:

> the most important mathematical propositions only achieve their simplest form on the introduction of this square root as a new symbol. Its justifiction thus rests in the propositions themselves. In a similar way the experience of present-day physics show us that atoms do not exist as simple material objects. However, only the introduction of the concept 'atom' makes possible a simple formulation of the laws governing all physical and chemical processes.[5]

Therefore, as Heisenberg points out, the new physics does not require that the classical concept of material atoms and *a fortiori* the classical concept of matter be discarded. Like the square root of -1, these concepts are useful, indeed necessary, when describing not only what we observe (our sense experiences) but also for our interpretations of those observations and for the laws (including the mathematical laws) we formulate. The concept of matter as a form of energy does not replace but enriches the older concept.

SUB-ATOMIC PARTICLES – RECAPITULATION

The theory of wave/particle duality gradually emerged as it became apparent that atoms were not, after all, ultimate particles. Sub-atomic particles were discovered in the late nineteenth century. First, there was Thomson's evidence for the existence of electrons; then later work in the early twentieth century revealed the existence of protons and neutrons; even more sub-atomic particles have since been detected. Second, it became apparent that some atoms were unstable and disintegrated spontaneously, emitting radiant energy and losing mass. Third, some of the newly discovered sub-atomic particles were found to be very very unstable and disintegrated quickly into radiant energy. These experimental findings were supported by Einstein's theory showing a mathematical relation between mass and energy. Matter has been shown to be a form of energy. In the quotation below, Heisenberg states that every form of energy has mass so that the converse is true and energy can be called a form of matter. For example, photons, that is light quanta, exemplify a form of energy but they can also be regarded as elementary particles:

> in considering three basic substances, i.e. three kinds of elementary particles – electrons, protons and neutrons – as the component parts of all matter, we have not altogether covered the programme of atomic physics. . . . If only these three elementary particles existed, we could rest satisfied in the belief that there are three fundamentally different sorts of matter which can no longer be transformed into one another or related to one another. But in reality there are yet other forms of matter, the most important being radiation. Since the famous formula of relativity theory has linked energy and mass, we know that every form of energy also possesses mass and that it can therefore be called a form of matter. According to Planck and Einstein, energy in radiation is concentrated in so-called light quanta which can also be regarded as some kind of elementary particles.

But beyond that, still other elementary particles have been found. ... [Yet] ... Our present concepts still seem to be too simple: there are many indications that further elementary particles exist which have not yet been observed because they have an extremely short life ... elementary particles can change into one another ... This is a process characteristic in general for the collision of two elementary particles of high energy.[6]

Bohr suggested that we should regard the wave and particles aspects of matter not as contradictory, but as complementary so that both were needed. Like Heisenberg he thought that matter and radiation were abstractions and, also like Heisenberg, he thought they were indispensable:

The individuality of the elementary electrical corpuscles is forced upon us by general evidence. Nevertheless, recent experience, above all the discovery of the selective reflection of electrons from metal crystals, requires the use of the wave theory superposition principle in accordance with the ideas of L. de Broglie. Just as in the case of light, we have consequently in the question of the nature of matter, so far as we adhere to classical concepts, to face an inevitable dilemma which has to be regarded as the very expression of experimental evidence. In fact here again we are not dealing with contradictory but with complementary pictures of phenomena, which only together offer a natural generalization of the classical mode of description. In the discussion of these questions, it must be kept in mind that, according to the view taken above, radiation in free space as well as isolated material particles are abstractions, their properties on the quantum theory being definable and observable only through their interaction with other systems. Nevertheless, these abstractions are, as we shall see, indispensable for a description of experience in connection with our ordinary space-time view.[7]

ELECTROMAGNETIC WAVES AS MATHEMATICAL FUNCTIONS

Like atoms and sub-atomic particles, electromagnetic waves are abstractions and the picture of them as consisting of transverse vibrations in a luminiferous ether must be regarded as purely metaphorical. This does have the great advantage of removing the problem of the nature of the ether, for there is now no necessity to postulate its existence. The electromagnetic waves are described in mathematical equations and can be treated as mathematical functions with no more reality than the square root of –1. It may be convenient to explain certain effects ('their interaction with other systems', see quotation above), such as interference and diffraction in terms of waves or indeed in terms of particles. Thus interference and diffraction patterns might be explained in terms of crests and troughs of waves or in terms of distribution of energy and, since matter is a form of energy, the pattern could be expressed as distribution of particles.

We have seen that diffraction is one characteristic of wave motion, and indeed the diffraction of light appeared to confirm it as a form of wave motion. It was shown that if light is passed through a narrow slit the rays will 'fan out' beyond it so that the slit acts as a fresh source of waves; there would be crests (areas of maximum illumination) and troughs (areas of darkness). But we have also seen that in certain circumstances light rays may be taken to be cannonades of photons and, when treated as such, the observed 'crests' and 'troughs' would represent maximum and minimum numbers of photons.

DIFFRACTION OF ONE PHOTON – A THOUGHT EXPERIMENT*

Observation and experimentation are intrinsic to the nature of empirical science and to indulge in thought experiments might seem to be the very antithesis of scientific practice. What can be the value of imagining the results of situations which cannot be confirmed (or refuted) by observation? But imaginative speculation is also intrinsic to scientific advance and when physicists are dealing with entities too small and/or of too low energy to be observed, it can be helpful to speculate as to how these might behave and to plan a thought experiment. Thought experiments will encourage theoretical speculation so that it may be hoped that ultimately empirical tests could be carried out to confirm or refute the results of such speculations.

Though it is not possible to isolate just one photon of light in practice, it is possible to isolate one photon in thought. We may imagine reducing the intensity of a ray of light shining on a screen with a fine slit and opening the slit for an extremely short time. Then it would, in theory, be possible for one and only one photon to pass through. With particle-like properties the photon would travel in a straight line, but since we cannot tell the angle at which it would pass through the slit, we cannot predict where it would go. This thought experiment is helpful because it demonstrates our inability to forecast what will happen: the direction and therefore the final position of a single emergent photon cannot be predicted, even in thought, i.e. even in principle.

PROBABILITY AND STATISTICAL PREDICTION

(a) Photons

But we *can* predict (and confirm by observation) the diffraction patterns to be obtained from ordinary rays of light consisting of very many photons. Max Born (1882–1970) suggested that diffraction patterns actually obtained when light is passed through a slit indicate the *probability* that any particular photon will go to a given place beyond the slit. The pattern shows light and dark bends (crests and troughs) and any photon is most likely to go to a crest position, where there is maximum light. It certainly will *not* go to a trough position, where there is darkness, but there *is* a possibility it will go somewhere in between the areas of darkness and those of maximum brightness. Since a diffraction pattern may be directly observed, or may be calculated knowing the wave length of the irradiating light and the size of the slit, it can be used to predict the *probable distribution* of the emergent photons. The consequence is that though we cannot predict the path of any particular photon, we can at least predict the *probability*, the statistical odds, for any given path.

(b) Electrons

We have seen that electrons can show diffraction patterns when beamed through slits formed by ordered atoms in a crystal. If the density of the electron beam is reduced, individual electrons will show a probability distribution analogous to that of individual photons. Just as with light, the direction of any particular electron cannot be predicted but, again, we can use the diffraction patterns to predict the probability of its taking any given path.

Born's suggestion of a probability distribution gives some sort of physical reality to the unpredictable behaviour of particles conceived as waves and of waves conceived as particles. Heisenberg wrote of the *tendency* of events to take place as having a kind of reality:

> The concept that events are not determined in a peremptory manner, but that the possibility or 'tendency' for an event to take place has a kind of reality – a certain intermediate layer of reality, halfway between the massive reality of matter and the intellectual reality of the idea or image – this concept plays a decisive role in Aristotle's philosophy. In modern quantum theory this concept takes on a new form; it is formulated quantitatively as probability and subject to mathematically expressible laws of nature.[8]

In that passage Heisenberg contrasts the 'massive reality of matter' with the 'intellectual reality of the idea' and this seems to echo the Platonic contrast between material objects and Universals. For Plato the material objects were but shadows, true reality inhered in the immaterial Universals accessible only to reason. For Heisenberg and Bohr electromagnetic waves and atomic and sub-atomic particles are symbols with no more ultimate reality than the square root of –1 whereas the mathematical equations, accessible only to reason, are ultimate reality. One cannot carry this analogy too far, but it does show that, in relation to matter, empirical science values accounts based on reason as highly as, if not higher than, those directly related to sense experience.

Statistical probabilities rather than determinant certainties illustrate one aspect of the study of matter which was unknown in classical physics. But, by appeals to tendency, we can reconcile this strange property of matter to our common-sense intuitions. However, we have to take the notion of indeterminancy further. Our thought experiments have been concerned with locating particles passing through slits; these showed that no unambiguous prediction could be made as to the behaviour, and therefore the position, of any given particle. But the indeterminancy as to position is more general than this.

In chapter 7 it was shown that there is uncertainty inherent in the dual nature of matter and it is shown in Schrödinger's treatment of particles as wave packets (see p. 127). As the position of a wave packet becomes more precisely determined, its energy (related to its wave length) becomes more difficult to calculate. In the next chapter we shall consider the fundamental uncertainties as to the properties of matter which our current theories appear to require.

SUMMARY

The classical distinction between matter and energy no longer exists; matter is a form of energy. Matter can be converted into energy and energy, in the form of electromagnetic waves, can be treated as matter.

The classical concept of atoms has had to be further modified since, in an ultimate sense, both particles and waves were best described in terms of mathematical equations.

Nevertheless, the classical concepts of waves and particles are useful as metaphors and should not be discarded. Bohr suggested that these two aspects of matter were to be regarded as complementary.

Diffraction patterns can then be interpreted either as representing crests and troughs of waves or as representing the presence and absence of particles.

Thought experiments show that it is not possible to predict the path and final position of individual electrons or photons after diffraction. Born suggested that diffraction patterns (involving many photons or electrons) indicate the probability that any given electron or photon will arrive at a given place. Thus though individual predictions cannot be given it is possible to make statistical predictions.

Schrödinger's exposition of wave packets shows that there is an inherent indeterminancy in measuring position and energy.

9

Problems

In classical physics there is, and always will be, a lack of precision in measurement. We cannot measure any quantity with absolute accuracy so any size, any weight, any velocity, any position, any time must always be stated within certain limits of error. Today we can make such measurements extremely accurately, but not with infinite accuracy. Yet this, though not entirely a practical problem, is practical to the extent that we can always aspire to achieve an accuracy which is sufficient for the purpose in hand. In theory we can go on refining our measurements indefinitely.

But there is another kind of imprecision which is inherent in all material measurement though it was first exhibited on the sub-atomic scale. Here we reach an uncertainty of a different kind: the more accurate the measurement, and therefore the more precise our knowledge of one quantity, the less accurate must be the measurement and therefore the less precise our knowledge of another quantity. The lack of precision, i.e. the degree of uncertainty, is not a result of any possible lack of technique, (that may be there as well, of course, but it can be ignored for our present purposes). It is inherent in our current theories as to the physical nature of matter.

One approach to an understanding of this characteristic is through a reconsideration of Schrödinger's wave packet. It has been shown that a wave packet, if built up from a sufficient number of waves, will have a single sharp crest in a precisely determinable position. But because then there is only *one* wave the concept of wave length (the distance from crest to crest, or trough to trough, etc.) is inapplicable. It follows that the concept of frequency is inapplicable and therefore the momentum and energy of the wave packet cannot be known. On the other hand, if we consider the component series of waves that constitute the wave packet, each will have measureable wave lengths, and so their frequency, momentum and energy can be known. But now there is no fixed position. Between these two limiting cases there will be bunches of several waves so that some estimate of wave length (therefore of frequency, momentum and energy) is possible. In addition, because of the 'bunching' some estimate of position can also be made. However, in such cases though neither momentum nor position can be *accurately* measured, there can be an approximate estimate of both, but it cannot be precise. Now wave packets are a way of describing the behaviour of sub-atomic particles, indeed wave packets correspond to material particles. So if particles are treated as wave packets there is an inbuilt uncertainty as to *their* momentum and position. Such was Bohr's argument. He was concerned to show that Schrödinger's account of matter in terms of wave packets along with the introduction of quantum theory, must make accurate knowledge of any physical system impossible.

Bohr also pointed out that there was another way of approaching the problem of inevitable uncertainty, in that the very process of observation must, of necessity, involve energy disturbance by the observer and therefore disruption of the system. This was also a feature of Heisenberg's exposition of his Uncertainty Principle.

HEISENBERG'S UNCERTAINTY PRINCIPLE

In 1927 Heisenberg published his account of a thought experiment whereby the inherent impossibility of precise measurement might be demonstrated. He considered how the position of an electron might be measured by observing it under a gamma-ray microscope*. In order to be observed, the rays reflected by the electron would have to to be focused through a lens (the microscope's objective*) and, because the lens constituted an opening, the rays would tend to be diffracted. For a sharp image (and therefore a defined position) diffraction must be negligible; this can be achieved using a large lens (i.e. a large opening) and gamma rays of very short wave length. However, if the illuminating rays are of very short wave length we know they will have a high frequency, a high energy and high momentum (see p. 123). High momentum rays will disturb the momentum of the electron. Hence, if we try to fix its position with short wave length rays we produce uncertainty of momentum, but if we reduce the disturbance, and the uncertainty of momentum, by using rays of longer wave length and less energy, diffraction effects will make the position of the electron uncertain.

Heisenberg showed that if the position of the electron is within a range Δx and the momentum is within a range Δp, then the product of the two must be greater than, or at the very best equal to, h (Planck's constant).

That is:

$$\Delta x \times \Delta p \geq h$$

Heisenberg also showed that a similar relationship of uncertainty existed in the establishing of the kinetic energy of a moving particle or wave packet at a definite time.

UNCERTAINTY AND THE DUALITY OF MATTER

Both these uncertainty relations are implicit in the concept of the wave packet where wave length (and therefore frequency and therefore energy) and position cannot both be measured accurately. Moreover, the uncertainty is not of a kind that might be converted into a statistical probability analogous to estimating the position of a photon. Heisenberg contrasted the statistical position in relation to the study of heat with the inherent uncertainty of quantum mechanics. In classical thermodynamics we have to make statistical predictions because we are ignorant of details of the system, for example the individual behaviour of literally trillions of molecules. In classical physics uncertain predictions 'can be taken to indicate an as yet unsolved problem'[1] and not necessarily one that is insoluble. But this is not the case in quantum mechanics for here 'the statement that an atom is in its normal state implies a complete knowledge of the atom concerned'.[2] Such definite statements make statistical statements irrelevant. The uncertainty in quantum mechanics is not a consequence of ignorance that we may hope to dispel, or at least reduce, but to an inescapable lack of knowledge.

There was some disagreement between Heisenberg and Bohr. Neither doubted the validity of the uncertainty relations, the issue was a difference as to the conceptual foundations of those relations.

Bohr related uncertainty to the wave/particle duality of matter, an indeterminacy which was well revealed in Schrödinger's account of wave packets. He thought the difficulties arose not because an account of matter in terms of particles and an account of matter in terms of waves or wave packets was misguided but because it was impossible to use both accounts together, 'in spite of the fact that only their combined use provides a full description of physical phenomena'.[3]

By contrast Heisenberg thought that indeterminacy was a consequence of mathematical and logical deduction from the equations and that therefore it was not necessary to appeal to wave/particle duality. Not that he wanted to dismiss the wave/particle picture entirely, but for him reality lay in the equations. We may remember that he regarded references to atoms and particles (see p. 131) as symbolic. He conceded that perhaps such references were indispensable but the reality was the mathematics. Bohr also appreciated that 'the symbolical garb of the [quantum-mechanical] methods ... closely corresponds to the fundamentally unvisualizable character of the problems concerned',[4] but he believed that wave/particle complementarity provided a real (and not *just* a symbolic) backing for the equations.

Bohr pointed out that Heisenberg's own derivation of indeterminacy relations from thought experiments was in fact based on the Einstein–de Broglie equations, which connect wave and particle descriptions and so implicitly presuppose wave/particle duality.[5] Even though eventually Heisenberg came to appreciate Bohr's point and the importance of Bohr's principle of complementarity. Heisenberg conceived indeterminacy as showing the limitations of classical concepts whereas Bohr sought some way of combining them.

Quantum physics showed that the concepts of classical physics, ultimately arising from direct observation, were inadequate. Many of those concepts – for example, atom, mass, velocity, momentum and energy – are sophisticated and have evolved slowly. Others – for example, particle, position, distance and time – are closely connected with common-sense intuition. Yet all physics must ultimately be based on observation and the new quantum physics must lead to further evolution of concepts analogous to the evolution of the sophisticated concepts of classical physics. Jammer maintains that 'deep in his heart' Heisenberg hoped that a reinterpretation of the concepts of position and velocity might resolve this problem so that it would become 'meaningless to speak of the place of a particle with a definite velocity'.[6] Heisenberg also began to approach Bohr's position in regard to the complementary relation of wave and particle concepts when he wrote, 'it is the simultaneous recourse to the particle picture and the wave picture that is necessary and sufficient to determine in all instances the limits to which classical concepts are applicable.'[7] This appears to be an attempt to develop the classical concepts so that they can be used in the newly discovered world of high-energy sub-atomic particles.

THE PRINCIPLE OF COMPLEMENTARITY AND THE LIMITATIONS OF CLASSICAL CONCEPTS

Bohr formulated his principle in several different ways[8] but essentially all formulations reflect his view of the duality of matter and the necessity to appeal both to wave and to particle descriptions in order to deal with matter in quantum mechanics. He wrote:

> this feature of complementarity is essential for a consistent interpretation of the quantum-theoretical methods. . . . Heisenberg . . . had pointed out the close connection between the limited applicability of mechanical concepts and the fact that any measurement which aims at tracing the motions of elementary particles introduces an unavoidable interference with the course of the phenomena and so includes an element of uncertainty which is determined by the magnitude of the quantum of action. This indeterminancy exhibits, indeed, a peculiar complementary character which prevents the simultaneous use of space-time concepts and the laws of conservation of energy and momentum, which is characteristic of the mechanical mode of description.[9]

The principle of complementarity reflects the limitation of classical concepts:

> The two views of the nature of light are [indeed] to be considered as different attempts at an interpretation of experimental evidence in which the limitation of the classical concepts is expressed in complementary ways.[10]

Jammer writes:

> in Bohr's view, the indeterminism of quantum mechanics has its origin in the unavoidable 'rupture' of description; for what Heisenberg called 'the reduction of the wave packet' was for Bohr, so to say, the switching over from one mode of description to its complementary mode.[11]

Bohr wanted to break away from the established notion that progress in research came through detailed analysis of phenomena. That methodology was the basis of classical physics and had been promulgated by Bacon, by Descartes, by Galileo and by Newton. But Bohr wanted to encourage a holistic approach.[12]

For Heisenberg, 'the particle picture and the wave picture are merely two different aspects of one and the same physical reality'.[13] He regarded energy as the one and only fundamental substance, but energy could exist in different discrete forms: 'It always appears in definite quanta which we consider the smallest indivisible units of all matter and which, for purely historical reasons, we do not call atoms but elementary particles.'[14] However, the particles of quantum physics do not have the properties of the particles of classical physics:

> we use symbols for the position and velocity of particles in Newtonian mechanics. If, however, we make use of any of the common symbols, such as the co-ordinates of a particle, we are already tacitly implying the existence of a given particles. Yet it is the decisive point of this last stage of atomic physics that particles can no longer be taken for granted ... thus we cannot sensibly assume co-ordinates and mass of definite particles. The question arises as to what we can use. We have not really developed the mathematical tools which would grasp these complex events ... is it really true that particles have position?[15]

INFLUENCE OF THE OBSERVER

We saw in note 9, above, that Bohr acknowledged that measurement introduced 'an unavoidable interference with the course of phenomena'.[16] Such interference is of negligible consequence in macro-measurements but becomes important in the miniature world of atoms and sub-atomic particles and so:

> Nature thus escapes accurate determination, in terms of our commonsense ideas, by an unavoidable disturbance which is part of every observation. It was originally the aim of science to describe nature as far as possible as it is, i.e. without our interference and our observation. We now realize that this is an unattainable goal. In atomic physics it is impossible to neglect the changes produced on the observed object by observation.[16]

Quantum mechanics differs from classical mechanics in that there is an inevitable interaction between the observer and what is observed. We cannot, as in measurements made in classical physics, assume that what is observed is independent of the process of observation. This is apparent in Heisenberg's thought experiment, given above, where he showed that the size of the lens aperture and the intensity of illumination must inevitably affect precision of measurement. Bohr himself acknowledged that 'an independent reality in the ordinary physical sense can neither be ascribed to the phenomena nor to the agencies of observation'. [17]

Heisenberg wrote that in actual (as opposed to thought) experiments there was a certain freedom to choose 'the dividing line between observer and what is observed'. Note that he took the physical state of the system to be mathematically defined:

> With the aid of more or less complicated apparatus we put questions to nature directed towards establishing some objective process in space and time. We may, for example, want to know whether electrons are deflected at a certain place. In this situation it follows automatically that, in a mathematical treatment of the process, a dividing line has to be drawn between, on the one hand, the apparatus which we use as an aid in putting the question and thus, in a way, treat as part of ourselves, and, on the other hand, the physical system we wish to investigate. The latter we represent mathematically as a wave function. This function, according to quantum theory, consists of a differential equation which determines any future state from the present state of the function. But we are satisfied with the laws formulated in terms of classical concepts for the making

of our apparatus and feel entitled to use them for measuring purposes. The dividing line between the system to be observed and the measuring apparatus is immediately defined by the nature of the problem but it obviously signifies no discontinuity of the physical process. For this reason there must, within certain limits, exist complete freedom in choosing the 'position' of the dividing line.[18]

Bohr related the inevitable influence of the observer to the consequent inability to apply causal laws:

the definition of the state of a physical system ... claims the elimination of all external disturbances.... if in order to make observation possible we permit certain interactions with suitable agencies of measurement ... an unambiguous definition of the state of the system is naturally no longer possible and there can be no question of causality in the ordinary sense of the word.[19]

CAUSAL LAWS

It is important to stress again that the Heisenberg uncertainty relations are not a consequence of clumsiness in measurement or lack of technique – we can assume zero clumsiness in thought experiments – rather, they are inherent in our concept of the nature of matter. In practice, clumsiness of measurement must *increase* the margin of error beyond the Heisenberg minimum. That minimum indicates that prediction on the basis of causal laws is also only possible within the limits established by Heisenberg's relationship.

In classical physics we establish causal relations in the form of laws (laws of nature) and then, from knowledge of the current state, we can predict future behaviour. We can predict the path of a billiard ball on a billiard table and the path of a capsule through space. In these cases, as we well know, there may be interference which upsets the predictions, but such interference does not show our causal laws are false, for the interference itself is subject to causal laws. Particularly good examples of the successful application of causal laws are predictions (and retrodictions) of the paths of heavenly bodies, because experience has shown that there is less chance of untoward events. So we confidently describe the appearance of the heavens for thousands of years in the future and thousands of years in the past. Even so, there is a limit to that time span (see below). If an unforeseen event did occur it would falsify the predictions but, as with the billiard ball and space capsule, it would not falsify the causal laws.

However, predictions based on the causal laws established in classical physics presuppose that the initial conditions (position, velocity, time, etc.) can be known, or rather that they can be measured as accurately as is required in a given context. It was this assumption that led various natural philosophers, such as Descartes and Laplace, to affirm that, at least in principle, it was possible to predict the future (and retrodict the past) indefinitely. The causal laws of classical physics are the basis of predictions, and in so far as the law is trustworthy, the prediction is trustworthy. This holds at least for relatively short-term predictions. But even on the macro-scale (and not considering quantum effects) the fact that there is an inevitable technical limit to accuracy makes it impossible to predict indefinitely into the future, even for astronomical events.

ULTIMATE INDETERMINANCY

But in the realm of sub-atomic entities, where it is impossible in principle to know both position and energy accurately, there can, in principle, be no accurate prediction. However it does not follow that there are no causal laws; it can be argued that they do exist, but that we cannot formulate them and that, even if we could formulate them, we would be unable to apply them. Moreover we have to concede that the predictability of macro-events is also limited by this uncertainty. The fact that for all practical purposes such predictions are reliable is a consequence of the very small value of the quantum of action, i.e. Planck's constant, h. At the macro-level the limitations inevitable with our measuring techniques are at present far greater than the theoretical quantum limitations. But should we come to achieve extremely great technical accuracy in measurement there would still be an inevitable quantum barrier. Bohr makes this point:

> any measurement which aims at an ordering of the elementary particles in time and space requires us to forego a strict

account of the exchange of energy and momentum between the particles and the measuring rods and clocks used as a reference system. Similarly any determination of the energy and momentum of the particles demands that we renounce their exact co-ordination in space and time. In both cases the invocation of classical ideas, necessitated by the very nature of measurement, is, beforehand, tantamount to a renunciation of a strictly causal description.[20]

We have either to accept that there is something inherently indeterminate about the physical world or that our present concepts of matter are inadequate. Heisenberg went so far as to reject the notion of causal laws: 'the incorrectness of the law of causality is a definitely established consequence of quantum mechanics itself.'[21] This was to take a positivist attitude to causality, in some ways analogous to Einstein's (and Newton's) opinion of the ether. Heisenberg conceded that there might be causal laws but affirmed that, since they could not be applied, it was pointless to speculate on that possibility:

> it is possible to ask whether there is still concealed behind the statistical universe of perception a 'true' universe in which the laws of causality would be valid. But such speculation seems to us to be without value and meaningless, for physics must confine itself to the description of the relationships between perceptions.[22]

By contrast Einstein, although he had taken a positivist view of the ether, never accepted that causal predictions could not be made, saying, in effect, that if this were the case he would rather be a croupier in a casino.[23] Was this an atavistic wish to retain something of classical physics and the classical concepts based on common-sense observation? In attempting to support his 'gut feeling' Einstein devised a series of thought experiments whereby he hoped to demonstrate to Bohr, and others, that quantum mechanics was inconsistent. His critical paper was published in 1935 and a report (not written by Einstein) was published in *The New York Times*, along with a non-technical explanation:

> Physicists believe that there exist real material things independent of our minds and theories. We construct theories and invent

words (such as electron, positron, &.) in an attempt to explain to ourselves what we know about our external world and to help us to obtain further knowledge of it. First the theory must enable us to calculate facts of nature, and these calculations must agree very accurately with observation and experiment. Second, we expect a satisfactory theory, as a good image of objective reality, to contain a counterpart for every element of the physical world. A theory satisfying the first requirement may be called a correct theory while, if it satisfies the second requirement, it may be called a complete theory.[24]

Since quantum theory did not give 'a good image of objective reality' because it did not 'contain a counterpart for every element of the physical world', Einstein believe that though it was correct as far as it went, it was incomplete. He was not alone in his views, for others agreed that unless it was assumed that reality was independent of us and our observations there was no common ground for discussion.[25]

There could then be a return to a phenomenalist position. Thus in 1971 Professor E.C. Kemble wrote:

the province of the physicist is not the study of an external world, but the study of a portion of the inner world of experience [and] there is no reason why the constructs introduced . . . need correspond to objective realities.[26]

Kemble wished to query Einstein's basic assumption that the sharp distinction between subject and object 'that serves us so well on the everyday large-scale common-sense level of discourse' could be sustained.[27] The debate is not yet over, but the tentative and generally agreed conclusion is that we have to accept uncertainty and, therefore, at the atomic level, we have to accept our inability to apply causal laws and to make predictions. It could also be the case that some of the constructs introduced, particularly mathematical functions, do not correspond to objective physical reality. But it does not follow that there is no ground for objective discussion and that therefore there must be an inevitable return to phenomenalism.

We tend to think of classical physics as the physics of intuitive judgements, but in fact this is not the case. Causal laws, that is the Cartesian 'laws of nature', did not arise from naive intuition but were the result of sophisticated theories and assumptions. Before the rise of classical physics teleological explanations in terms of the intentions and purposes of gods and animate beings were the norm. Classical physics, the physics of Galileo, Descartes and Newton represented a break and a development from the latent animism and the common-sense postulates of Aristotelian physics. Likewise classical corpuscular and atomic theories and the emergence of the concept of elements were an advance from intuitive assumptions about the nature of the material world. In all these cases original thinkers from Thales to Dalton made bold postulates which at the time seemed counter-intuitive. It is only because the new postulates became part of accepted explanations that they were incorporated into common-sense knowledge. As science progressed they were refined and in part replaced by later developments in classical physics and chemistry.

Both for educated common sense and for classical physics atoms represented ultimate reality but, if we follow Heisenberg, quantum mechanics has a new reality in mathematics. Observed events remain real enough; in one sense this is a return to pre-classical ideas, but atoms and sub-atomic particles are mathematical entities:

> In the experiments about atomic events we have to do with things and facts, with phenomena that are just as real as any phenomena in daily life. But the atoms or the elementary particles themselves are not as real; they form a world of potentialities or possibilities rather than one of things or facts.[28]

And Bohr himself acknowledged the fundamentally abstract nature of matter:

> There is no quantum world. There is only an abstract quantum physical description. It is wrong to think that the task of physics is to find out how nature *is*. Physics concerns what we can say about nature.[29]

It should be noted that this is not to return to phenomenalism in that there can be objective agreement as to what we can say about nature.

It may be that we shall come to replace physical descriptions of micro-events by equations, and that we shall abandon appeals to causal laws and accept as a fact that there is ultimate indeterminancy and uncertainty. Undoubtedly this will bring about a new common-sense concept of explanation. What are the philosophical implications that this might bring to the problem of matter?

SUMMARY

Research at the sub-atomic level reveals an inherent indeterminancy in the properties of matter because the more accurate the measurement of one quantity the less accurate must be the measurement of another quantity. For example, it is impossible, even in an ideal situation with no fault in technique, to obtain precise measurements of both position and momentum. The limit is set by Planck's constant in that the product of the uncertainty of position and the uncertainty of momentum must be greater than, or at best equal to h, the quantum of action.

This uncertainty stems from the dual nature of matter and also from the inevitable interaction between the observer and the phenomena. It applies to all measurement but, because Planck's constant is so small, it is only significant at the sub-atomic level.

Classical concepts of energy and position are inadequate to describe sub-atomic events. Bohr's principle of complementarity acknowledges that both the wave and the particle aspects of matter must be appealed to; the principle reflects the limitation of classical concepts.

The inability to make precise measurements makes it impossible to specify an original state precisely and therefore impossible to apply causal laws of nature. It may be that there are such laws but this is of little interest since they can be of no use to us. On the other hand, there may not be any ultimate causal laws so that there is an ultimate indeterminancy in nature. This conclusion was resisted by Einstein, who believed that further research would lead to a more developed theory and that the present quantum theory was incomplete.

Some who accept indeterminancy are resigned to a subjective phenomenalist position but objective accounts need not be ruled out by quantum theory. Nevertheless, it may be that we shall have to accept that accounts of sub-atomic entities and events can only be given in the form of mathematical equations and that physics cannot tell us what nature *is*. This would entail a new philosophical assessment of the problem of matter.

10

Recapitulation and Assessment

MATTER, MATERIALS AND PHYSICAL OBJECTS

Physical objects are everywhere; they include our own bodies, the bodies of other animals, plants and an indefinitely large number of materials and inanimate objects. We know that all these exist because many of their properties are evident to sense perception. Sense perception, direct or indirect, provides our only evidence, and it is the ultimate basis for all our claims to knowledge of the world. But there have been speculations as to the nature of the physical world which go beyond what can be sensed. We have seen that during the past 2,500 years there has been a steady development of theories as to the underlying nature of materials and physical objects, theories which have evolved from the sophisticated speculation that there is some fundamental substance which supports, or is in some sense responsible for, the varying sensible qualities we perceive. Jammer suggests that attempts to explain the visible world in terms of an invisible world incorporate the notion of 'hidden variables', i.e. invisible entities, and that the idea of hidden variables 'is as old as physical thought'.[1] As we saw in chapter 1, claims to explain the visible in terms of the invisible were embodied in other theories from the sixth century BC. Some of these were speculations as to there being an underlying homogeneous substance and others postulated the existence of tiny separate and indivisible entities. This is matter.

CONCEPTS OF MATTER

As has been repeatedly stressed, the concept of matter is a sophisticated abstraction based on inferences from what is directly observed, but until the twentieth century it was generally assumed that though matter was invisible it was potentially visible and tangible. The hidden variables, whatever their nature, were regarded as concrete material entities. For some there was one fundamental substance which was closely related to familiar materials, for example water or air. Others suggested a small number of different substances, the elements, which were also closely related to familiar materials. Heraclitus suggested fire which he conceived as a principle of change: fire was the unity underlying change.

(a) Supra-sensible Reality

Parmenides and Plato did indeed think of ultimate reality as being non-sensible and only to be reached by thought. We may be tempted to relate their ideas and those of Heraclitus to twentieth-century scientific theories of energy expressed in terms of mathematical equations, but this would be misleading. Plato believed that knowledge of the Universals was arrived at by thought, not observation: for him ultimate reality *transcended* observation, it was not dependent on it. Plato's appeal to mathematics as providing a subsidiary criterion of knowledge was based on his view of mathematics as a branch of pure reason. He did not think that the impermanent physical objects and materials were proper objects of knowledge though, influenced by Pythagoras, he did relate their properties to mathematical/geometric structures. However this is a far cry from modern accounts of hidden variables (sub-atomic entities) in terms of differential equations. Likewise Heraclitus did not have the classical physicist's concept of energy, so he did not think of heat (fire) as a *form* of energy. *A fortiori* he did not conceive of matter as energy.

Nearly 2,000 years later Leibniz took the fundamental *materia prima* to be a collection of immaterial souls. These were *his* hidden variables. Unlike the early Greek conceptions of matter they were indeed immaterial and, moreover, they were activated by a force, *entelechy*. But Leibniz's entelechy was analogous to a living or vital force. The latent animism in his account and his justification by appeal to metaphysical theories distinguishes it sharply from twentieth-century theories. Only Boscovich's treatment of atoms as points of force can be convincingly related to the twentieth-century view; indeed, Popper (p. 32) regarded

Boscovich's theory as a forerunner of the nineteenth-century field of force.

(b) Atomic Theories

These early philosophical speculations were important in the development of the concept of matter but, in relation to current scientific ideas the atomos, corpuscular and atomic theories were especially influential. Of necessity these theories were of a hypothetical character for all the atomists accepted that the tiny corpuscles were too small to be perceived. But from the time of Democritus, there were always attempts to show that their existence could be inferred from what was perceived and there were attempts to use them to explain those properties and changes in properties which could be directly perceived. Moreover,

> Though atoms represented the final indivisible 'brick' of matter, they nevertheless appeared to be miniature parts of ordinary matter. The atom then, at least in our imagination, was endowed with all the macroscopic properties of matter.[2]

Eighteenth- and nineteenth-century advances in experimental chemistry and chemical theory suggested that different types of atoms were responsible for the properties of different elements.

(c) Phenomenalism

Atomic theories were not universally accepted and, during the nineteenth century, phenomenalists proposed a totally different approach to the problem of matter *and* of material objects. They asserted that the concept of matter, or substance, as an underlying reality was superfluous and that all references to materials and material objects could be re-expressed in terms of actual or possible sense experiences. In chapter 5 we saw that though this approach does indeed dispose of the problem of matter, simply by denying its existence, it offers a sterile solution. The baby is thrown out with the bath water and we no longer have any way of explaining *how* or *why* we have sense experiences or why we can claim to transcend solipsism to have objective knowledge of the world.

(d) Divisible Atoms

Towards the end of the nineteenth century advances in experimental science and observational techniques revealed some of the previously hidden variables in so far as they could be directly inferred from observation. It became possible to investigate the structure of atoms and to study their behaviour in electric and magnetic fields. Hence all doubts as to the reality of atoms were dispelled. But it became apparent that atoms, which had been thought to be the fundamental units of matter, were themselves divisible; they were no longer to be conceived as immutable. In one sense this made for a more satisfying account because, instead of having to postulate a fairly large number of different atoms for each of the elements, it became possible to envisage basic matter as consisting of a smaller number of ultimate particles which could be built up to make the different atoms of the different elements. Thus, at the start of the twentieth century matter was still regarded as being an agglomeration of tiny material particles. Atoms were no longer basic but were themselves conceived as minute mechanical systems of sub-atomic particles; they were likened to miniature solar systems.

PROBLEMS

But there were some difficulties with this mechanical picture. It had been assumed that the energy of the system would change smoothly so that there would be a continuum of energy states. Yet such an assumption was incompatible with the sharp spectral lines obtained from activated atoms. The anomaly was dispelled by Planck's discovery that there was an essential element of discontinuity in any description of atomic events. There were ultimate and indivisible quanta of action; hence action, and therefore energy, were not continuously varying quantities. Indeed Planck's quantum theory shows that there is this same discontinuity in all events but on the macroscale it is far too small to be observed or to be of any practical consequence.

There were further dramatic developments, as experimental observations showed that tiny sub-atomic particles could, in certain circumstances, behave like electromagnetic waves, waves of the same nature as light. Moreover, at the same time, it was shown that light waves, photons, could be treated as particles. Wave/ particles had strange properties revealed by quantum theory in that their momentum and position could not be precisely measured. This property of indeterminism is inherent in all matter so that there is some imprecision in measurements of visible particles, such as grains of sand and of physical objects, such as tables and chairs. But, just as with energy discontinuities, the imprecision is far too small to be detected or to be of any consequence.

In addition Einstein's theory of relativity showed that material particles could no longer be regarded as being essentially corpo-real since *matter* was a form of energy.

LIMITATIONS OF SENSE PERCEPTION AND CONCEPTS

All these developments revealed a fundamental limitation of human perception: our senses cannot register such tiny changes of energy and position. Because of this limitation we form concepts of a world in which matter and energy are continuous and in which the position and energy of physical objects can be precisely measured. Nevertheless there have been major modifications in that from early times, and *a fortiori* in classical physics, one naive common-sense concept was revised. It became accepted that matter was discontinuous and was built up from discrete particles. It was acknowledged that atoms could not be seen, but since they were potentially visualisable they could be accommodated to our common-sense concept of physical objects.

However our common-sense concepts and also the concepts of classical physics, still presuppose a world where precise measurements are possible. This is a world where we can apply causal laws, laws which are a basis for reliable predictions. Indeed, in all but special circumstances, this is still the case since, because our senses are too gross to detect its effects, in ordinary observations we neglect the discrete nature of the quantum of action. Bohr has argued that our ordinary perceptions should be regarded as *idealisations* whereby we make order of our sense impressions. That order rests on the fact that, within the limits of our unaided senses, light has, in effect, an infinite velocity and the quantum of action is, in effect, negligibly small.[3] We have to accept these instinctive idealisations as our inevitable starting point. They are based on our sense experiences and they colour our whole language, for all our sense experience must ultimately be expressed in language.

Heisenberg points out that, with a comparatively moderate demand for accuracy (albeit quite considerable in relation to daily experience), it is not nonsensical to speak of the position and velocity of an electron.[4]

CONFLICT WITH COMMON-SENSE INTUITION

Quantum theory and relativity theory lead us to conclude that the apparent stability of physical objects is an illusion and that the underlying substance of materials and material objects is itself immaterial. This is a counter-intuitive notion and such a bald assertion can strike us not only as just counter-intuitive but as absurd. However, if we consider the way we have arrived at such a view of matter, though it may still remain strange and counter-intuitive it can no longer be regarded as absurd.

Our previous three chapters have shown how these new concepts of physical objects and of matter arose. They describe how scientific study, that is experimental observation, inferences from observation and imaginative conjectures show the nature of matter to be far removed from the common-sense picture and far more mysterious than earlier philosophical and scientific speculations ever suggested. Matter has been shown to be a form of energy and in that form it has mass. Yet, in bulk, matter supports the tangible material of physical objects and materials that we sense by touch and by sight. As Bohr said:

> only by a conscious resignation of our usual demands for visu-alisation and causality was it possible to make Planck's discovery fruitful in explaining the properties of elements on the basis of our knowledge of the building stones of atoms.[5]

SUB-ATOMIC ENTITIES AS MATHEMATICAL ABSTRACTIONS

We may agree with Heisenberg that, though it is convenient to refer to particles, and indeed to sub-atomic entities as particles, these have little more than a metaphorical status; they exist only as variable in mathematical equations. So an alternative approach to understanding the new concepts of matter is to avoid specuations as to the nature of tiny particles and to treat them as symbols for variables in quantum equations. We cannot visualise the space–time and quantum worlds and the knowledge we have of sub-atomic entities is inferred from experiments on radiation and collision reactions. The inferences we make, whereby we interpret the actual experimental evidence, are helped by appeal to mathematical equations and to thought experiments. These are intellectual abstractions but, as Bohr says, 'our whole space–time view of physical phenomena, as well as the definition of energy and momentum, depends ultimately upon these abstractions'.[6]

THE NECESSITY FOR RECONCILIATION

Such a concept of matter as a form of energy and/or as a set of symbols in equations is indeed difficult to reconcile with our common-sense intuitions. Science has made matter yet more mysterious and it seems as if we are abandoning any hope of describing the miniature world of atoms. Yet, if we are to make any sense of our scientific theories we must try to give an account of the mystery in terms of the concepts and language we use to account for ordinary observations:

> It lies in the nature of physical observation . . . that all experience must ultimately be expressed in terms of classical concepts, neglecting the quantum of action.[7]

It is not only that we aspire to understand the quantum world as something more than a mathematical abstraction, we need to relate quantum events to experimental observations and these are of necessity given in terms of ordinary concepts. Though from the times of the development of classical physics our scientific theories have embodied mathematical abstractions, these have been related to experimental observations. It is here that the concepts of common sense and classical physics are essential for they are the only means we have to describe our experiences, i.e. to describe what we observe:

> The necessity of using the language of classical physics when discussing observational facts followed, for Bohr from our

inability to forgo our usual forms of perception; it also makes it unlikely that the fundamental notions of classical physics could ever be abandoned for the description of physical experience.[8]

Some physicists insist that it is possible and indeed useful to treat sub-atomic entities as more than mathematical abstractions and to conceive them as with at least some particulate properties. Thus Born said, 'Though an electron does not behave like a grain of sand in every respect, it has enough invariant properties to be regarded as just as real.'[9]

But others, though conceding that experimental observations can and must be described, believe that ultimately quantum entities are mere symbols (see also p. 131). Heisenberg said that we cannot carry out the original aim of science, namely to describe nature as it is, for such a description cannot be given in terms of our common-sense ideas.[10] Thus we use classical concepts to describe our experiments and we can give factual descriptions of these but the theories that *explain* our observations are abstractions. Heisenberg said:

In the experiments about atomic events we have to do with things and facts, with phenomena that are just as real as any phenomena in daily life. But the atoms or the elementary particles themselves are not as real; they form a world of potentialities or possibilities rather than one of things or facts.[11]

As we saw in chapter 9, Bohr took the same view and said;

There is no quantum world. There is only an abstract quantum physical description. It is wrong to think that the task of physics is to find out how nature is. Physics concerns what we can say about nature.[12]

THE REFINEMENT OF LANGUAGE AND CONCEPTS

What we can say may have to be in the language of mathematics but we may not have to give up all hope of at least sketching quantum events with words.

Language is intimately involved with thought, and the psychological difficulties we have in arriving at appropriate concepts are reflected in and reflected by the limitations of our language. We must look to evolve new concepts and a new language. Thus Heisenberg asserts that though we must use our existing terms to describe our experiences, those terms must be revised in the light of new experiences.[13] We might also say that they must be revised in the light of new theories and that such revision had started with classical physics long before the advent of quantum theory. Already there has been reference to the incorporation of atomic theories into a concept of matter. In addition the classical concepts of inertia, of momentum and of energy themselves depend on abstraction from direct experience. Heisenberg says that 'even the mathematically exact sections of physics represent . . . only tentative efforts to find our way among a wealth of phenomena'.[14] Those classical terms represent abstractions, but they refer to refinements of what is observed and, once understood, they are not counter-intuitive. Even though they are inadequate they must be used if we are to give some physical content to our equations:

> It was the increased range of technical experience which first forced us to leave the limits of classical concepts. These concepts no longer fitted nature as we had come to know it. We observed the track of an electron moving as a particle . . . and, on another occasion, we found it reflected . . . like a wave. The language of classical physics was no longer capable of expressing these two observations as effects of a single entity.[15]

Bohr suggested that our capacity to create descriptive concepts rested on our differentiation between subject and object. He stressed that science presupposes objectivity and aims to avoid all reference to the observer. Experimental results should be valid as objective truths and should be quite independent of the observer and the process of observation. Mathematical symbolism sets up an ideal of objectivity.[16]

THE LOSS OF OBJECTIVITY

Bohr thought that since the meaning of a concept or word depends on an arbitrary choice of viewpoint, we must accept that a full elucidation of one and the same object may require 'diverse points of view which defy unique description'.[17] He pointed out that our concepts of space and time have their meaning because we ignore our interaction with the means of measurement.[18] Yet in dealing with sub-atomic events:

> any attempt at an ordering in space–time leads to a break in the causal chain since such an attempt is bound up with an essential exchange of momentum and energy between individuals and the measuring rods and clocks used for observation; and just this exchange cannot be taken into account if the measuring instruments are to fulfil their purpose.[19]

Our psychological picture of the ordinary world rests on a sharp distinction between space and time and is a consequence of the smallness of the velocities of ordinary objects as compared with the velocity of light. Similarly, the fact that we find that a causal space–time description is suitable for ordinary phenomena depends on the smallness of the quantum of action.[20] In everyday life and in macro-scientific investigations we can and do separate and distinguish the observer from the process of observation. But in the quantum world this is not possible.

Therefore, in the quantum world we have to abandon the Baconian approach of investigation through analysis even though it was the Baconian principle of dissection and analysis which revealed that world. The difficulties which have emerged demand the adoption of a more holistic approach.[21] We must now include ourselves, the observers, in our pictures of the world:

> The new situation in physics is that we are both onlookers and actors in the great drama of existence.[22]

QUANTUM PHYSICS AND CONSCIOUS AWARENESS

Bohr suggested that there were reciprocal relations which depend upon the unity of our consciousness and which exhibit a striking similarity with the physical consequences of the quantum of action, for example, the well-known characteristics of emotion and volition which cannot be represented visually. He contrasted the outward flow of associative thinking and the preservation of the unity of personality and compared this with the contrast between wave and particle descriptions. He suggested that the unavoidable influence of atomic phenomena caused by observing them perhaps corresponded to the change of the tinge of psychological experiences which accompanies any direction of the attention to one of their various elements.[23]

Bohr also compared the problem of lack of causality in atomic phenomena to the problem of free will. The deterministic picture of a causal chain in physical brain processes although not at present practicable would indicate a unique chain of physical reactions (governed by laws of cause and effect). But if we try to relate physical processes in the brain to psychical experiences, we must be prepared to accept that observation of the processes will in itself bring about an essential alteration in conscious awareness. Similarly, detailed causal tracing of atomic processes is impossible since any attempt at observation of necessity involves interference in the course of those processes. The analogy is acceptable if are observing our *own* brain events but it does not follow that an observer of another person's brain events would influence that other's conscious awareness.

Bohr thought that perhaps 'in the facts which are revealed to us by the quantum theory we have acquired a means of elucidating general philosophical problems'.[24] This may or may not be the case, but at least we can say that there is an analogy between the limitations of our ordinary concepts of physical events insofar as describing the miniature quantum world is concerned and the limitation of our concepts of physical processes insofar as we relate them to our own consciousness.[25]

MATTER AND MIND

If we surmise that all forms of mental activity are a result of the activity of a physical brain and associated nerve tissue, we must then regard consciousness and conscious awareness as an emergent property of matter. It would seem that matter, organised in certain particular ways (in our experience organic compounds ordered in certain living cells), becomes conscious. There is nothing especially surprising in this for we see many properties emerging as organisation increases. For example, the properties of water emerge as a result of a certain organisation of hydrogen and oxygen atoms, and the properties of polythene emerge as a result of a certain organisation of carbon and hydrogen atoms. Organisation of carbon, hydrogen, oxygen, nitrogen and the atoms of a few other elements can produce living tissue. Materialist theories of mind depend on acknowledging the possibility of emerging properties in organized matter.

Our current scientific theories tell us that all matter, including of course brains and nervous tissue, is essentially a form of energy, non-tangible and non-visualisable. Those scientific theories may change, but they are unlikely to take us back to a view of matter as consisting of ultimate particles. Hence any materialist theory of mind must relate consciousness to energy and not to the corporeal entities postulated in earlier forms of materialism. As Armstrong has said:

> a physicalist philosophy is not at the end, but rather at the beginning, of its problems. The clearing away of the problem of mind only brings us face to face with the deeper problems connected with matter.... A physicalist theory of mind is a mere prolegomenon to a physicalist metaphysics. Such a metaphysics ... will no doubt be the joint product of scientific investigation and philosophical reflection.[26]

THE NEED FOR REASSESSMENT

The new knowledge from quantum physics has shaken the foundations underlying the building up of concepts on which not only the classical description of physics rests but also our ordinary mode of thinking.[27] Just as developments in classical physics have modified naive intuitions and concepts, so developments in quantum physics may bring about further modifications. We need to reassess our world, including ourselves and other living creatures, in the light of the new concept of matter. This will influence more than scientific theories for it will also, as Armstrong says, alter the significance of philosophical materialism.

But that is a topic for another book. Here we must acknowledge that matter remains a mystery and a problem.

SUMMARY

Matter is an abstraction from the sensible world of materials and physical objects.

There have been two kinds of different concepts of matter, firstly as being some form of supra-sensible reality and secondly as being constituted of hidden variables: tiny ultimate particles, atoms, too small to be seen.

There has also been the view that the concept of matter should be discarded and replaced by reports of sense experiences but this is a sterile solution to the problem of the nature of matter.

In the twentieth century atoms were shown to be divisible; they were not ultimate particles.

Quantum theory and relativity theory showed that subatomic entities had properties which seemed very different from those of macro-matter. So different were they that it was suggested that references to particles (or to waves) were no more than useful metaphors and that the hidden variables were best regarded as mathematical abstractions. It was also shown that matter could be represented in a non-corporeal form as energy.

It became clear that macro-matter had the same properties as sub-atomic entities and that these properties had not been detected because our sense perceptions were too crude. Our common-sense concepts of physical objects had taken no cognisance of them just because those concepts were based on our limited sense perceptions. In one sense, the concepts might be called idealisations.

Yet we have to rely on our sense perceptions and on the concepts arising from them. They are the basis of our language and therefore the means by which we describe observations, including experimental observations. We must try to refine our language to accommodate quantum physics just as common-sense language was refined to accommodate classical physics.

Quantum theory also shows that there can be no clear-cut distinction between the observer and what is observed. Bohr suggested that there was an an analogy between observation of subatomic events and observation of conscious awareness since, in this latter case also, there is no clear-cut separation of consciousness and brain events.

It seems necessary that we reassess and develop our concepts of matter in order to make further progress in the philosophy of mind as well as in scientific research.

Notes

PREFACE

1. B. Russell, *A Critical Exposition of the Philosophy of Leibniz*, George Allen & Unwin, London, 1975, p. 75.
2. A.J. Ayer, 'I think, therefore I am', *Descartes*, ed. W. Doney, Macmillan, London, 1970, p. 83.
3. B. Russell, *The Problems of Philosophy*, Oxford University Press, Oxford, 1974, p. 10.
4. In this book the world 'metaphysical' signifies non-empirical and can also refer to those basic assumptions which must underpin scientific (empirical) theories. See J. Trusted, *Physics and Metaphysics*, Routledge, London, 1991, pp. ix–xi.
5. Russell, *The Problems of Philosophy*, p. 91.

1 EARLY THEORIES

1. Aristotle distinguished planets from stars: he thought the former did not twinkle because they were near. Nevertheless it was quite usual to use the word 'star' to refer to all the heavenly bodies apart from the sun and moon.
2. From the seventeenth century onwards, the distinction between animate and inanimate matter became sharper but, in the twentieth century, the study of viruses has again made the boundaries more hazy.
3. We must bear in mind that such participation did not necessarily imply personal gods and goddesses (see also note 10).
4. Andrew G. Van Melsen, *From Atomos to Atom*, Harper & Brothers, New York, 1960, p. 10. In fact, only two of the words cited ('theory' and 'hypothesis') derive directly from Greek. The other examples are translations from Latin. But those Latin words are themselves translated from the Greek of Aristotle and later writers.
5. G.E.R. Lloyd, *Early Greek Science. Thales to Aristotle*, Chatto and Windus, London, 1970, p. 19.
6. G.S. Kirk, J.E. Raven and M. Schofield, *The Presocratic Philosophers*, 2nd edn., Cambridge University Press, 1983, pp. 90–1.
7. op. cit., pp. 93–4.
8. op. cit., p. 101.
9. op. cit., pp. 96–7.
10. W.K.C. Guthrie, *The Greek Philosophers. From Thales to Aristotle*, Methuen, London, 1967, p. 32.
11. Kirk *et al.*, op. cit., p. 110.
12. op. cit., p. 119.

13. Lloyd, op. cit., p. 21.
14. Guthrie, op. cit., p. 29.
15. Kirk *et al.*, op. cit., pp. 145–6
16. Lloyd, op. cit., p. 37.
17. Guthrie, op. cit., p. 50.
18. Kirk *et al.*, op. cit., p. 198.
19. Cyril Bailey, *The Greek Atomists and Epicurus*, Oxford, 1928, p. 19.
20. By the fifth century BC the Greeks were able to extract iron from its ores. The Iron Age was later in Northern Europe.
21. Kirk *et al.*, op. cit., p. 358.
22. op. cit., p. 368.
23. Lloyd, op. cit., p. 40.
24. Georgius Agricola, 'On the Origin of Metals', from *De Ortu et Causis Subterraneorum*, 1546; trans. Herbert Clark Hoover and Lou Henry Hoover, 1912; from Georgius Agricola, *De Re Metallica*, 1556. New York Dover reprint, 1950, p. 51. Also in *Science In Europe 1500–1800 a Primary Sources Anthology*, ed. Colin A. Russell, Open University, 1991, p. 49.
25. Aristotle in Lloyd, op. cit., p. 25.
26. Guthrie, op. cit., pp. 14–5.
27. Werner Hesenberg, trans F.C. Hayes, *Philosophic Problems of Nuclear Science*, Faber and Faber, London, 1952, p. 28.
28. Simplicius in Kirk *et al.*, op. cit., p. 426.
29. Democritus, quoted by Heisenberg, op. cit., p. 97.
30. Ibid.
31. Guthrie, op. cit., p. 58.
32. Heisenberg, op. cit., p. 30.
33. Van Melsen, op. cit., p. 31.
34. Today we tend to think of materials and objects as being nothing more than a collection of their qualities; we may appeal to molecular structure but the notion of substance seems redundant. We might also suggest that there seem to be some qualities, like the blue of the sky, which are not carried by any material substance.
35. Note that qualities *per se* had to inhere in a substance; form was a more abstract conception, a principle of being.
36. But matter, though structureless and unknowable, could itself be the source of individuality and difference. If the form of two or more things were identical (for example, two newly minted pounds) they were to be distinguished by their matter. The property of individuation was especially important in the doctrine of transubstantiation since the priest was able, with divine and miraculous aid, to change the matter of bread and wine into the matter of the flesh and blood of Christ. In this case the form (and therefore the appearance) remained unaltered.
37. Note that primary matter differs from substance in that, at least in theory, substance was the *independent* support for qualities.
38. D.J. O'Connor, 'Aristotle', in *A Critical History of Western Philosophy*, ed. D.J. O'Connor, Macmillan, Inc. New York and Collier-Macmillan Publishers, London, 1964, p. 50.

39. In this respect Aristotle's view of elements differed from that of Empedocles.

2 NON-MATERIAL REALITY

1. It is not clear whether Plato thought natural objects like mountains were or were not permanent; in early times, and indeed up to the seventeenth century AD, they were thought to be unchanging. If Plato thought otherwise it would have been for theoretical rather than empirical reasons.
2. Guthrie maintains that, after Socrates, Parmenides was the greatest single influence on Plato. It was Parmenides 'that giant of intellect among the Presocratics whose challenging thesis that by all rational argument motion and change were impossible had to be met without evading his apparently unassailable premises. W.K.C. Guthrie, *A History of Greek Philosophy*, Volume IV, Cambridge University Press, 1975, p. 34.
3. Plato, *The Complete Texts of Great Dialogues of Plato* ed. and trans. W.H.D. Rouse, Plume Books, New York, 1961, pp. 371–6.
4. G.E.R. Lloyd, *Early Greek Science, Thales to Aristotle*, Chatto and Windus, London, 1970, p. 71.
5. Lloyd, op. cit., pp. 71–2.
6. Regular solids have all faces of equal size and shape.
7. Lloyd, op. cit., p. 74.
8. Guthrie writes: 'The very word *philosophia* as Plato uses it is a link between them [the Pythagoreans] ... his passion for mathematics as a glimpse of eternal truth ... his choice of musical terminology ... [etc.] ... are evidence of a close affinity between the two in which Plato must have been a debtor.' Guthrie, op. cit., p. 35.
9. Werner Heisenberg, trans. F.C. Hayes, *Philosophical Problems of Nuclear Science*, Faber and Faber Ltd., London, 1952, p. 33.
10. Plato said that the regular solids were themselves built from plane surfaces; he showed that these surfaces were made up of right-angled isosceles triangles. The triangles could be arranged as squares of different sizes corresponding to three grades of the element earth (Lloyd, op. cit., p. 76).
11. Lloyd, op. cit., pp. 76–7.
12. 'Plato's relations with Democritus are a fascinating but tantalizing subject, for he never mentions him, yet it is impossible to believe that he was not acquainted with his work or that, if acquainted, he did not react strongly.... Democritus called his ultimate realities *ideai*, though for him this denoted millions of irregularly-shaped, solid physical atoms. These ultimate realities were beyond 'bastard cognition' of the senses and, like Platonic Forms, were accessible only to thought. This made him a more dangerous foe, but foe he remained, for he committed the ultimate blasphemy of denying purpose in the universe and teaching a soulless, irrational mechanism.' Guthrie, op. cit., p. 37.
13. Heisenberg, op. cit., p. 98.

14. Lloyd, op. cit., p. 72.
15. Lloyd, op. cit., p. 73.
16. L.J. Russell, *'Leibniz'*, *Encyclopaedia of Philosophy*, Vol. 4, Crowell Collier and Macmillan Inc., New York, 1967, p. 426.
17. ibid.
18. ibid.
19. G.H.R. Parkinson, ed. G.H.R. Parkinson and Mary Morris, trans., *Leibniz. Philosophical Writings*, Dent, London, 1984, pp. 116–17.
20. L.J. Russell, op. cit., p. 426.
21. We could better describe this force as energy.
22. Bertrand Russell. *The Philosophy of Leibniz*, Allen and Unwin, London, 1975, pp. 105–6.
23. Leibniz did not think of mind as brain and therefore did not entertain the possibility of the mind having different parts.
24. Parkinson, op. cit., p. 118.
25. L.J. Russell, op. cit., p. 429.
26. Gerd Buchdahl, 'The Interaction Between Science, Philosophy and Theology in the Thought of Leibniz', *Studia Leibnitiana Symposium* of the Leibniz-Gesellschaft, Reading, July 1979, p. 75. But not all commentators agree, see Bertrand Russell, op. cit., pp. 149–50.
27. C.D. Broad, *Leibniz, An Introduction*, Cambridge University Press, 1975, p. 107.
28. op. cit., p. 108.
29. Buchdahl, op. cit., p. 80.
30. Parkinson, op. cit., p. 92.
31. op. cit., pp. 181–2.
32. Bertrand Russell, op. cit., pp. 121–5.
33. L.J. Russell, op. cit., p. 430.
34. R.J. Boscovich, *A Theory of Natural Philosophy*, ed. and trans. J.M. Child, MIT Press, 1966, p. 19.
35. op. cit., p. 160.
36. op. cit., p. 63.
37. op. cit., p. 21.
38. op. cit., pp. 20–1.
39. op. cit., p. 65.
40. op. cit., pp. 25–6.
41. K.R. Popper, 'Philosophy and Physics', *The Myth of the Framework*, Routledge, London, 1994, p. 116.
42. ibid.
43. op. cit., p. 117.
44. Boscovich, op. cit., p. 67.
45. L. Pearce Williams, *Michael Faraday*, London, 1965, pp. 72–9.
46. Heisenberg, op. cit., pp. 55–6.
47. Immanuel Kant, *Critique of Pure Reason*, trans. J.M.D. Maiklejohn, Everyman, London, 1934, p. 196.
48. ibid.
49. Kant, op. cit., p. 144.
50. Kant, op. cit., p. 157.
51. Kant, op. cit., p. 162.

3 FROM ATOMOS TO CORPUSCLES

1. Some philosophers distinguished soul and spirit from mind but many, including Descartes, postulated just one *immaterial* substance.
2. Descartes, *Philosophical Writings*, eds. E. Anscombe and P. Geach, Nelson, The Open University, 1971, p. 207.
3. Anscombe and Geach, op. cit., XXIII, p. 208.
4. Anscombe and Geach, op. cit., p. 199.
5. Anscombe and Geach, op. cit., p. 234.
6. Anscombe and Geach, op. cit., CXCVII, p. 232.
7. Anscombe and Geach, op. cit., VI, p. 200.
8. Anscombe and Geach, op. cit., p. 205.
9. Anscombe and Geach, op. cit., p. 206.
10. Bernard Williams, *Descartes*, Penguin, 1978, p. 230.
11. Williams, op. cit., p. 239.
12. Anscombe and Geach, op. cit., p. 225.
13. Descartes' analysis of force and motion was also shown to be incorrect by Leibniz.
14. Anscombe and Geach, op. cit., XLV, p. 224.
15. Today we take these 'laws' to be descriptive rather than prescriptive; they are *our* ideas, not God's.
16. Anscombe and Geach, op. cit., pp. 236–7.
17. Descartes thought that animals (the brutes) were solely corporeal, they had no soul (spirit); God and the angels were pure spirit; only human beings had a dual nature.
18. There are physical *events* in the human body, for example digestion, and these are distinguished from bodily *actions* which are, or are thought to be, controlled by conscious thought.
19. At first Cartesian dualism seems a most reasonable way of accounting for human nature, but there are insurmountable difficulties in explaining interaction of mind and body.
20. For examples *The Self and Its Brain*, by K.R. Popper and J.C. Eccles, Springer International, London and New York, 1977.
21. A. Koyré, *Metaphysics and Measurement*, Chapman and Hall, London, p. 122.
22. R. Boyle, 'The Origin of Forms and Qualities According to the Corpuscular Hypothesis' in *Works*, Vol. III, Johnston, Crowder, Payne *et al.*, 1777, p. 15.
23. ibid.
24. Boyle, op. cit., p. 16.
25. Note the Cartesian dualism assumed here.
26. Boyle, op. cit., p. 23.
27. Boyle, op. cit., p. 29.
28. Here is an early introduction of the concept of molecules.
29. Mercuric oxide, a new compound, though readily decomposed back to mercury and oxygen. It was with mercuric oxide that Priestley isolated oxygen in the eighteenth century.
30. Mercury vapour, the result of a physical change of state, not chemical action.

31. Boyle, op. cit., p. 30.
32. Boyle, op. cit., p. 34.
33. Boyle, op. cit., p. 35.
34. Locke was a natural philosopher but is here disassociating himself from experimental investigation; a hint of what was to become the separation of philosophy from natural science.
35. Here Locke uses the term 'philosophy' to include natural science.
36. J. Locke, *An Essay Concerning Human Understanding*, IV, 3, xvi.
37. Locke, op. cit., II, 23, xxvi.
38. Locke, op. cit., II, 23, ix.
39. R. Jackson, 'Locke's Primary and Secondary Qualities', in *Locke and Berkeley*, eds. C. B. Martin and D.M. Armstrong, Macmillan, London, p. 60.
40. Locke, op. cit., II, 8, ix–xv.
41. Newton's prestige was so great that the wave theory was not revived until the early nineteenth century when Thomas Young argued it gave better explanations of various phenomena. In the twentieth century corpuscular and wave theories are integrated. See chapter 8.
42. F. Greenaway, *John Dalton and the Atom*, Heinemann, London, 1966, p. 28; quotation from Newton's *Principia*.
43. Arnold Thackray, *Atoms and Powers*, Harvard University Press, Cambridge, Mass., 1970, p. 22.
44. Isaac Newton, *Opticks*, 1704, Query 31, Quoted by Thackray, op. cit., p. 23.
45. Joseph Priestley; quoted by Thackray, op. cit., p. 54.
46. Thackray, op. cit., p. 146.

4 FROM CORPUSCLES TO ATOMS AND MOLECULES

1. For example, explanations of thunder and lightening (p. 12), physical laws (p. 46) and chemical change (p. 29) can be related to the corpuscular theories of Democritus, Boyle and Boscovich.
2. Corpuscular theories were rightly called philosophical theories when 'philosophy' signified 'natural philosophy' and embraced natural science. But today philosophy is more or less divorced from science and if we use the term 'philosophical theory' the current implication is that the theory is metaphysical and non-empirical.
3. Popper originally argued that though observation could not conclusively establish a theory as true, it could show that it was false. See K.R. Popper, *The Logic of Scientific Discovery*, Hutchinson, London, 1972, p. 41. Despite Popper's own qualifications this is not now accepted.
4. These were proposed and developed throughout the eighteenth century.
5. It is interesting that what seemed to be experimental confirmation of the law of conservation of mass was almost contemporary with Einstein's work showing that mass could be converted into energy. This helped lead to a fundamental reappraisal of the concept of matter (see chapter 7). However, in ordinary chemical changes the

energy released produces changes in mass which are even less than one in 10,000,000, so the experimental investigation was not faulty, it was simply not sufficiently accurate. See also note 24.

6. Combustion can also occur without oxygen, for example, in the presence of chlorine, but combustion in air always involves oxygen.

7. In the early seventeenth century Francis Bacon (1561–1626) had concluded, on the basis of empirical observations, that heat was connected with motion, but the notion of caloric in the form of minute particles had superseded Bacon's theory.

8. John Dalton, *A New System of Chemical Philosophy*, Peter Owen, London, 1965, p. 1.

9. This is an example of another early tentative notion of the concept of molecules. See note 28, chapter 3.

10. Frank Greenaway, *John Dalton and the Atom*, Heinemann, London, 1966, p. 21.

11. Aristotle was supposed to have lectured in the covered walk-way, the *peripatos* of his school at Athens and Aristotle and his followers were called the peripatetic philosophers or, with Boyle's spelling, the peripateticks.

12. Robert Boyle, quoted by Rupert Hall, *The Scientific Revolution 1500–1800*, Longmans, London, 1967, p. 325.

13. Hall, op. cit., pp. 324–5.

14. Greenaway, op. cit., p. 23.

15. Thus Lavoisier listed heat, caloric, in his table of elements.

16. Newton's theory of gravitational attraction postulates an attractive force between all masses, inversely proportional to the square of the distance between them. This would hold for corpuscles as much as for the heavenly bodies.

17. Greenaway, op. cit., p. 30. Dalton held that heat caused repulsion. (See quotation 14.)

18. W.G. Palmer, *A History of Valency to 1930*, Cambridge University Press, 1965, p. 5.

19. ibid.

20. This law states that the weight of a substance A, which combines with a given weight of another substance B, will also combine with that weight of substance C which combines with the same given weight of B; or it will be a simple multiple (or fraction) of that weight.

21. In Dalton's time atomic weights were given as the ratio of the atom's weight to the weight of a single hydrogen atom. This was changed later; see chapter 6, note 20.

22. Dalton, op. cit., p. 167.

23. In fact water, H_2O, contains two hydrogen atoms so the atomic weight of oxygen (as calculated by Dalton) is half the correct ratio.

24. This is the law of conservation of mass which was adopted as an *a priori* principle in the late eighteenth century (see p. 61) and held until Einstein showed the relation of mass and energy ($E = mc^2$); the law of conservation of mass is still applied for ordinary chemical reactions where changes in energy are too small to produce detectable changes in mass. See also note 5.

25. Dalton, op. cit., pp. 162–3.
26. Greenaway, op. cit., p. 150.
27. W.H. Wollaston, 'On Super-acid and Sub-acid Salts', *Philosophical Transactions of the Royal Society of London*, 1808, p. 101.
28. Wollaston, op. cit., p. 102.
29. W.H. Wollaston, 'On the Elementary Particles of Certain Crystals', *Phil. Trans.*, 1813, pp. 54–6.
30. W.H. Wollaston, 'A Synoptic Scale of Chemical Equivalents', *Phil. Trans.*, 1814, p. 7.
31. Greenaway, op. cit., pp. 188–91.
32. It might seem that Crookes anticipated the modern concept of isotopes, with different numbers of neutrons in the nucleus, but this is not the case; he had no idea of a complex atomic structure.
33. Greenaway, op. cit., pp. 206–7.
34. They were usually aqueous solutions but sometimes the fused salt was electrolysed.
35. Palmer, op. cit., p. 23.
36. Greenaway, op. cit., pp. 206–7.
37. The gases had to be at the same temperature and pressure.
38. Assuming Avogadro's law the molecular weight of gas molecules can be calculated from the experimentally determined density (the vapour density of the gas). When we are dealing with an element, where all the atoms in each molecule are identical, then if we know the number of atoms in the molecule, the atomic weight of each atom is easily calculated. It is clear that apparently anomalous results will be obtained if changes in temperature produce a change in the number of atoms in the molecule.
39. The kinetic theory of gases explains the properties of gases on the basis that they are composed of molecules which are relatively far apart and which are moving with an average root mean square velocity which increases with temperature.
40. J.C. Maxwell, 'Atom', *Encyclopaedia Britannica*, 9th edn, 1875, Adam and Charles Black, Edinburgh, p. 40.
41. Maxwell, op. cit., p. 38.
42. Maxwell, op. cit., p. 41.
43. Maxwell, op. cit., p. 45.
44. ibid.
45. Greenaway, op. cit., p. 221.
46. Thomson had suggested that atoms of matter were vortex rings in the ether, like smoke rings in the air. 'The vortex was stable, it could oscillate, it might even form an association with another vortex to form something analogous to a molecule.' Greenaway, op. cit., p. 221. But on this basis it was difficult to explain the weight and density of material substances. See S.F. Mason, *A History of the Sciences*, Routledge and Kegan Paul, London, 1953, p. 392.
47. J.A.R. Newlands, quoted by Greenaway, op. cit., p. 218.
48. The valency of an element is in effect its combining power. But see Glossary.
49. There were exceptions but these will not be discussed here.

50. Compounds with the same chemical formula but with different structures (and therefore different properties) are *isomers*.
51. Optical isomers have identical chemical properties but opposite effects on polarised light. Pasteur showed that crystals of a pair of isomers were mirror images but he did not relate this to molecular structure.
52. Instrumentalists treat scientific theories as a potentially useful means of co-ordinating experimental data but they do not commit themselves to judging their truth. They regard this as irrelevant.

5 PHENOMENALISM

1. The term 'physical object' is defined as an entity which exists independently of sense experience.
2. Instrumentalism: scientific theories are not held to give descriptions, but are to be used solely as instruments for prediction and possible control of events. See also chapter 4, note 52.
3. These may also be called, *ideas*, percepts or sense data.
4. John Locke, *An Essay Concerning Human Understanding*, II, 23, ii.
5. George Berkeley, *Of the Principles of Human Knowledge*, XLIX.
6. God's *ideas* are relatively permanent, in the same way as physical objects are relatively permanent.
7. Quoted by A. Flew, *An Introduction to Western Philosophy*, Thames and Hudson, London, 1971, p. 343.
8. David Hume, *A Treatise of Human Nature*, ed. L.A. Selby Bigge, Oxford, 1973, Book I, Part IV, Section II, pp. 187–8.
9. Hume, op. cit., p. 198.
10. Hume, op. cit., p. 212.
11. Hume, op. cit., p. 213.
12. Hume, op. cit., p. 215.
13. Hume, op. cit., p. 218.
14. ibid.
15. John Stuart Mill, *Philosophy of Scientific Method*, Hafner Publishing Co., New York, 1950, pp. 44–5.
16. Mill, op. cit., pp. 371–2.
17. Ernst Mach, 'The Economical Nature of Physical Inquiry', in *Philosophy of Science*, ed. J.J. Kockelmans, Collier-Macmillan, London, The Free Press, New York, 1968, p. 180.
18. Mach, op. cit., p. 181.
19. ibid.
20. Max Jammer, *Concepts of Mass*, Harvard University Press, Cambridge, Mass., 1961, p. 100.
21. Mach, op. cit., p. 185.
22. ibid.
23. Karl Pearson, 'Perceptual and Conceptual Space', in Kockelmans, op. cit., pp. 200–1.
24. Émile Meyerson, 'Identity of Thought and Nature as the Final Goal of Science', in Kockelmans, op. cit., p. 345.

25. Meyerson, op. cit., pp. 345–6.
26. A.J. Ayer, *The Problem of Knowledge*, Pelican, 1956, p. 119.
27. Ayer, op. cit., pp. 126–7.
28. Ayer, op. cit., pp. 127–9.
29. Ayer, op. cit., p. 132.
30. Bertrand Russell, 'The World of Physics and the World of Sense', in Kockelmans, op. cit., p. 392.
31. Russell, op. cit., p. 393.

6 THE DIVISIBLE ATOM

1. Jennifer Trusted, *Beliefs and Biology*, Macmillan, London, 1996, p. 111.
2. Stephen Mason, *A History of the Sciences*, Routledge and Kegan Paul, London, 1953, p. 370.
3. Quoted in W.G. Palmer, *A History of the Concept of Valency to 1930*, Cambridge University Press, 1965, p. 127.
4. Quoted in Palmer, op. cit., p. 127.
5. Frank Greenaway, *John Dalton and the Atom*, Heinemann, London, 1966, p. 221.
6. Palmer, op. cit., p. 128.
7. Quoted by Mason, op. cit., p. 445.
8. Quoted by Isobel Falconer 'Corpuscles, Electrons and Cathode Rays: J.J. Thomson and the Discovery of the Electron', *British Journal for the History of Science*, 1987, 20, p. 259.
9. If two conducting plates differ in electric charge (one might be positive and one negative, or they could both be positive or both be negative, but to a different degree) then a potential difference (P.D.) exists between them. If they were to be connected by a conductor, a current would flow through the conductor. If they are separated by a low-pressure gas, then the electrical strain may produce the cathode ray effects described.
10. See later p. 104 and Glossary.
11. Quoted by Russell Stannard and Noel G. Coley in 'Introduction to Quantum Theory', Open University A381, *Modern Physics and Problems of Knowledge*, Block IV, Open University Press, Milton Keynes, 1981, p. 77.
12. Falconer, op. cit., p. 267.
13. See Jennifer Trusted, *Physics and Metaphysics*, Routledge, London, 1991, pp. 150–8. Ether was originally thought to be a very tenuous fluid permeating space; in the nineteenth century it was held to be theoretically necessary for the transmission of light waves. See also chapter 7.
14. Falconer, op. cit., p. 269.
15. Falconer, op. cit., p. 273.
16. Quoted by Falconer, op. cit., p. 273.
17. An atom or group of atoms can become ionised if they acquire a net positive (or negative) electrical charge. This will occur, for example, if electrons are lost or gained from a neutral atom or group.

18. Helium atoms have a double positive charge balanced by two electrons. If these electrons are lost, the atoms are ionised (see note 17 above) and carry a double positive charge.
19. Because atomic weight is related to atomic number Mendeleef's classification was not much disturbed. In fact, the new criterion removed the anomalies Mendeleef had found. (See Chapter 4, p. 72.)
20. Oxygen itself is a mixture of isotopes, but the predominant one has an atomic weight of 16 and this is now used as the standard.
21. Neils Bohr, *Atomic Theory and the Description of Nature*, Cambridge University Press, 1934, p. 26.
22. Bohr, op. cit., p. 27.
23. Bohr, op. cit., pp. 40–1.
24. Translated and quoted by Palmer, op. cit., p. 130.

7 THE DUALITY OF MATTER

1. Niels Bohr, *Atomic Theory and the Description of Nature*, Cambridge University Press, 1934, p. 31.
2. As early as 1900, Max Planck (1858–1947) had suggested that matter might exhibit particle-like properties in some circumstances and wave-like properties in others. This suggestion was developed by Einstein in his explanation of the photoelectric effect in terms of photons. See p. 121.
3. Newton developed the concept of inertia from its formulation by Galileo and Descartes but it was, of course, known to crude common sense that a force was needed to start or stop bodies moving and to change their direction.
4. As we saw in Chapter 2, Boscovich was an exception in that he did not regard the ultimate constituents of matter as material particles. We shall return to consider his view in chapter 8.
5. Werner Heisenberg, trans. F.C. Hayes, *Philosophical Problems of Nuclear Science*, Faber and Faber, London, 1952, p. 33.
6. Quoted by Jennifer Trusted, *Physics and Metaphysics*, Routledge, London, 1991, p. 158.
7. Quoted in Trusted, op. cit., p. 171.
8. Quoted in Trusted, op. cit., p. 152.
9. A particle might be deflected by hitting the edge of the slit, but would tend to reflect back on its path.
10. *Action* denotes the overall motion of a physical system; in relation to electromagnetic radiation, action multiplied by frequency gives a unit of energy. Planck asserted that there was a minimum unit of action denoted by h so that an electromagnetic wave of frequency f, would possess energy hf and this could not be subdivided.
11. The value of Planck's constant has been found to be $6{,}626 \times 10^{-34}$ joule-seconds.
12. Quoted by Russell Stannard and Noel G. Coley in 'An Introduction to Quantum Theory', Open University A382 *Modern Physics and Problems of Knowledge*, Block IV, Open University Press, Milton Keynes, 1981, p. 74.

13. Quoted by Stannard and Coley, op. cit., p. 73.
14. Heisenberg, op. cit., p. 14.
15. Heisenberg, op. cit., p. 46.

8 MATTER AND ENERGY

1. Even the UHF (ultra high frequency) and VHF (very high frequency) wireless waves have very long wave length (and low frequency) compared to heat and light.
2. Einstein arrived as his equation:

$$E = mc^2$$

Energy equals the product of mass and the square of the velocity of light from consideration of problems relating to the electrodynamics of moving bodies and his theory of relativity. See Jennifer Trusted, *Physics and Metaphysics*, Routledge, London 1991, pp. 171–5.
3. Werner Heisenberg, trans. F.C. Hayes, *Philosophical Problems of Nuclear Science*, Faber and Faber, London 1952, p. 51.
4. Heisenberg, op. cit., pp. 55–6.
5. Heisenberg, op. cit., p. 56.
6. Heisenberg, op. cit., pp. 102–3.
7. Neils Bohr, *Atomic Theory and the Description of Nature*, Cambridge University Press, 1934, pp. 56–7.
8. Quoted by Max Jammer, *The Philosophy of Quantum Mechanics*, John Wiley & Sons, New York, London, Sydney, Toronto, 1974, p. 44.

9 PROBLEMS

1. Werner Heisenberg, trans. F.C. Hayes, *Philosophical Problems of Nuclear Science*, Faber and Faber, London, 1952, p. 51.
2. ibid.
3. Max Jammer, *The Philosophy of Quantum Mechanics*, John Wiley & Sons, New York, London, Sydney, Toronto, 1974, pp. 65–6.
4. Neils Bohr, *Atomic Theory and the Description of Nature*, Cambridge University Press, 1934, p. 12.
5. Jammer, op. cit., p. 69.
6. Jammer, op. cit., p. 58. See also quotation 15, directly from Heisenberg.
7. Quoted by Jammer, op. cit., p. 59.
8. See discussion in Jammer, pp. 102–3.
9. Bohr, op. cit., p. 11.
10. Bohr, quoted in Jammer, op. cit., p. 92.
11. Jammer, op. cit., p. 98.
12. Jammer, op. cit., p. 199.
13. Heisenberg, quoted by Jammer, op. cit., p. 68.
14. Heisenberg, op. cit., p. 103.
15. Heisenberg, op. cit., pp. 104–5.

16. Heisenberg, op. cit., p. 73.
17. Bohr, op. cit., p. 64.
18. Heisenberg, op. cit., p. 49.
19. Bohr, quoted by Jammer, op. cit., p. 87.
20. Bohr, op. cit., p. 114.
21. Heisenberg, quoted by Jammer, op. cit., p. 75.
22. Heisenberg, quoted by Jammer, op. cit., p. 76.
23. Jammer, op. cit., p. 124.
24. Jammer, op. cit., p. 189.
25. Quoted by Jammer, op. cit., p. 193.
26. Quoted by Jammer, op. cit., p. 193.
27. ibid., p. 194.
28. Heisenberg, quoted by Jammer, op. cit., p. 205.
29. Bohr, quoted by Jammer, op. cit., p. 204. See also note 16.

10 RECAPITULATION AND ASSESSMENT

1. Max Jammer, *The Philosophy of Quantum Mechanics*, John Wiley & Sons, New York, London, Sydney, Toronto, 1974, p. 257.
2. Werner Heisenberg, trans., F.C. Hayes, *Philosophical Problems of Nuclear Science*, Faber and Faber, London, 1952, p. 72.
3. Niels Bohr, *Atomic Theory and the Description of Nature*, Cambridge University Press, 1934, p. 5.
4. Heisenberg, op. cit., p. 73.
5. Bohr, op. cit., p. 108.
6. Bohr, op. cit., p. 77.
7. Bohr, op. cit., pp. 94–5.
8. Jammer, op. cit., p. 100.
9. Born, quoted by Jammer, op. cit., p. 163.
10. Heisenberg, op. cit., p. 73.
11. Heisenberg, quoted by Jammer, op. cit., p. 205.
12. Bohr, quoted by Jammer, op. cit., p. 204. See also note 29 in Chapter 9.
13. Heisenberg, op. cit., p. 43.
14. Heisenberg, op. cit., p. 44.
15. Heisenberg, op. cit., p. 46.
16. Bohr, op. cit., p. 97.
17. Bohr, op. cit., p. 96.
18. Bohr, op. cit., p. 99.
19. Bohr, op. cit., p. 98. See also note 20 in chapter 9.
20. ibid.
21. Jammer, op. cit., p. 199.
22. Bohr, op. cit., p. 119. See also note 12 in chapter 9.
23. Bohr, op. cit., pp. 99–100.
24. Bohr, op. cit., p. 101.
25. ibid.
26. D.M. Armstrong, *A Materialist Theory of the Mind*, Routledge and Kegan Paul, London, New York, reprint 1976, p. 366.
27. Bohr, op. cit., p. 101.

Glossary

For page references in the text, see the Index.
The explanations given apply to the terms *as used in the text*; in some cases there are further meanings, not relevant here.

Action   an abstract quantity which describes the overall motion of a physical system. It can be thought of as the average momentum of the system multiplied by the length of the path between the initial and final position. It can also be thought of in terms of the average kinetic energy of the system and the time between reaching the final position from the initial position.

Alkali   *see* Base.

Animaliculae   microscopically small living creatures. When microscopes were first used in the seventeenth century all animate entities were termed animaliculae. Today we would class many of them as bacteria.

Animism   the view that all parts of matter involve consciousness so that there is no sharp distinction between animate and inanimate matter.

Base   a chemical compound which accepts hydrogen ions; alkalies are bases that are soluble in water.

Black body   a physical object with a dull black surface. Such bodies radiate and absorb heat more efficiently than bodies with any other kind of surface.

Calcination   an early term denoting combustion and heating.

Caloric theory of heat   an early theory of heat as a material substance; usually caloric was thought to be a liquid.

Calx   a substance formed after heating; in the eighteenth century an oxide of a metal was called a calx.

Cartesian dualism   the philosophical thesis, expounded by Descartes, that mind and body are two separate and distinct kinds of substances. The mind was equivalent to the soul and each human person was essentially a mind (soul). In life each mind was associated with a body but after death the mind (soul) would exist independently.

Change of state   the three states of matter are solid, liquid and gas. Melting, vaporising, solidification (freezing) and condensing are examples of changes of state.

Compound   a substance composed of two or more elements chemically combined.

Diffraction   the bending of light, or other waves, into the region of the geometrical shadow of an obstacle. The effect is especially noticeable with waves passing through apertures narrow in comparison with their wave length. The aperture effectively acts as a second source of waves.

Dispersion   the separating of white light into the rainbow colours.

Dualism   any philosophical theory which postulates two and only two different kinds of basic substances. It is to be contrasted with monism and pluralism. Cartesian dualism (see above) is one kind of dualism.

Electrode   an electrical conductor through which an electric current enters or leaves a conducting medium.

Electrolysis   a means of producing chemical changes through reactions at electrodes in contact with an electrolyte, by the passage of an electric current. The term often refers to the decomposition of aqueous solutions of acids, bases and salts.

Electrolyte   a chemical compound which when fused or dissolved in certain solvents, usually water, will conduct an electric current.

Element   a simple substance which cannot be chemically decomposed into more simple component parts. It is to be contrasted with a compound.

Empirical   applied to beliefs and theories which can be tested by appeal to observation (sense experience); they may thereby be confirmed or disconfirmed.

Energy   the capacity for moving mass, that is for overcoming inertia and for doing work.

Equivalent weight   the number of parts by weight of an element which will combine with 1,008 parts by weight of hydrogen, 8.00 parts by weight of oxygen and the equivalent weight of any other element or compound. Equivalent weights of different elements either react one to one or in a simple ratio.

Field of force   a domain in which gravitational, magnetic or electrical forces operate.

Frame of reference   a base to which to refer physical events. The position of any body can be established only in relation to another body (or bodies); in a laboratory it is convenient to refer the position to co-ordinate axes (x, y and sometimes z axes) as used in graphs. These axes can be taken as frames of reference.

Frequency   as applied to wave motion: the number of waves in a given unit of time, usually in 1 second.

Gamma-ray microscope   a microscope focusing gamma rays rather than light. Since these rays are of much shorter wave length than visible light, the microscope can detect objects too small to disturb (and therefore to be detected by) light waves.

Helmholtz Vortex ring   *see* Vortex Theory

Huygens' wave theory of light   the theory that light consists of longitudinal wave motion.

Inertia   the universal property of all physical mass in virtue of which it opposes any attempt to change its velocity.

Instrumental view of theories   the view that a scientific theory is to be regarded as an instrument for producing new predictions or for co-ordinating observational data. On this view a theory can be useful or not useful but cannot be said to be true or false.

Interference   the interaction between two or more sets of waves *where their path cross*. The resultant effect is the algebraic sum of the individual waves.

Ion an atom or group of atoms which by loss or gain of one (or more) electron(s) has acquired an electric charge.

Isomerism two or more compounds with the same chemical formula (that is, the same number and the same proportions of different atoms) but with different structures, are isomers and show isomerism. They usually have different chemical properties (but see also optical isomerism).

Isotope one member of a chemical element family of atoms all of which have a nucleus with the same number of protons but different numbers of neutrons.

Kinetic energy a form of energy possessed by a moving mass in virtue of its motion.

Law of equivalent proportions the weight of two elements, A and B, which combine separately with identical weights of another element, C, are either the weights in which A and B combine or are related to them in a small whole-number ratio.

Metaphysics originally signifying 'beyond physics' and concerning non-empirical matters, including religious and mystical beliefs. Today metaphysics may be concerned with assessing the presuppositions and assumptions that must form the basis for scientific knowledge.

Momentum the product of the mass and velocity of a moving body.

Monism the philosophical thesis that there is only one kind of substance. Monists may be materialists or idealists, or they may think that there is a primal material that is neither material nor spiritual (neutral monism), but which combines aspects of both the material and non-material. Monism is to be contrasted with dualism and pluralism.

Necessary truth a necessary truth is one that could not have been otherwise; it is sometimes called a logical truth.

Neutralisation the reaction between an acid and a base to produce a salt and water.

Neutron an uncharged sub-atomic particle found in the nucleus. Effectively, it has the same weight as a proton.

Noumenon (noumenal) Kant used this term to denote things as they are in themselves (*ding an sich*) as opposed to things known through sense-perception. He postulated the noumenal world as lying behind the phenomenal world of sense perception but affirmed we could know nothing about it. See also 'Phenomenon'.

Objective lens the convex lens in a microscope, near to the object observed; it is to be contrasted with the eyepiece through which the observer looks.

Optical isomerism optical isomers (see above) have different effects on polarised light but otherwise have the same chemical properties.

Phenomenon (Kant) Kant's term for the objects and events known (experienced through sense perception. They are constructed by us in accordance with our cognitions so as to appear extended in space and time and to show causal relations.

Photon an elementary light, particle *and* an informal unit of light energy (= hf). It gives a description of light in terms of energy packets rather than in terms of waves.

Plenum the conception of space as being completely full of matter; in a plenum there can be no vacuum.

Pluralism the philosophical thesis that there are more than two different kinds of primary substance.

Positivism the philosophical thesis that knowledge can only be acquired through sense experience. Metaphysics is rejected.

Potassium nitrate (nitre) a chemical compound of potassium, nitrogen and oxygen which readily ignites and is used in gunpowder.

Proton a positively charged hydrogen ion; protons are also a constituents of atomic nuclei.

Refraction the bending of light waves as they pass from one medium, into another, e.g. from air into water.

Solipsism the philosophical belief that only oneself and one's experiences exist.

Specific gravity the density of a material relative to the density of water.

Spectral lines elements and compounds, when activated by heat, vibrate so as to absorb electromagnetic energy from the spectrum. This effect is marked by dark bands (or lines) in the spectrum. Each substance shows characteristic spectral lines and may be identified by these.

Spectrum the range of electromagnetic waves from gamma rays through to wireless waves. This is the complete spectrum; in some contexts the term refers just to the visible spectrum, i.e. the rainbow colours.

Thought experiment an imagined experiment whereby the scientist may surmise that some result will follow. This may lead to a development or to a questioning of a current empirical theory.

Valency the combining power of an element whereby it reacts chemically with other elements. It may be measured by the number of links to other atoms that one atom of an element forms upon chemical combinations. Some elements show variable valency in different chemical combinations.

Velocity the distance a moving body travels in a given unit of time *in a given straight-line direction*. Velocity will change if speed or direction is changed.

Vortex theory; Vortex ring a vortex is a circular flow of liquid. Vortex theory derives from the Maxwellian view that electricity is a strained state of the ether and that electrons might consist of vortices in the ether.

Index